Laurent Schwartz

Cancer – Between Glycolysis and Physical Constraint

Springer

Berlin
Heidelberg
New York
Hong Kong
London
Milan
Paris
Tokyo

04 BoiU

Laurent Schwartz

Cancer – Between Glycolysis and Physical Constraint

Springer

Laurent Schwartz MD, PhD
Service de radiothérapie-cancérologie
Hôpital de la Pitié-Salpétrière
73 Boulevard de l'hôpital
75013 Paris
France
e-mail: laurent.schwartz@polytechnique.fr

ISBN 3-540-20496-2 Springer-Verlag Berlin Heidelberg New York

Library of Congress Cataloging-in-Publication Data

Bibliographic information published by Die Deutsche Bibliothek
Die Deutsche Bibliothek lists this publication in the Deutsche Nationalbibliografie; detailed bibliographic data is available in the Internet at http://dnb.ddb.de

Springer-Verlag is a part of Springer Science+Business Media

springeronline.com

Production: ProEdit GmbH, 69126 Heidelberg, Germany
Cover: Erich Kirchner, Heidelberg, Germany
Typesetting: K. Detzner, 67346 Speyer, Germany

Printed on acid-free paper 21/3150ML 5 4 3 2 1 0

To Laurence,
Madeleine, Leon, Jules
and Jeanne

Foreword

All great painters failed to be recognized in their lifetime. I'm a painter who has not been recognized in his lifetime therefore I'm great. In fact I am a painter, a gifted amateur, but with sufficient insight to know that I will never amount to much. The assertion in the first sentence is known as a syllogism and is a well-recognized error in logic. The same logical trap neatly describes the history of science. I have just finished reading the hugely entertaining book *A Short History of Nearly Everything* by Bill Bryson.[1] In this book he traces the history of the great discoveries in cosmology, geology, palaeontology and life science. Within this book he celebrates the giants and geniuses of scientific ideas since the age of enlightenment. It is certainly true that progress was often dependent on the revolutionaries who challenged the prevailing dogma and were in return ridiculed, dying in obscurity, before their ideas were accepted by the mainstream. Major shifts in our understanding of nature and the cosmos resulted from the overturning of the conceptual model of the problem rather than the accumulation of "facts".[2] But beware, it does not follow that all revolutionaries who are prepared to challenge the prevailing dogma are great scientists. The book that describes the other side of this coin is called *Voodoo Science* by Robert Park[3], equally entertaining but

[1] A short history of nearly everything, Bill Bryson, BCA 2003
[2] The Structure of scientific revolutions, T. Kuhn, London 1962
[3] Voodoo Science, Robert Park, Oxford University Press, 2000

sounding a cautionary note. Some self-appointed scientific revolutionaries can be both mad and bad!

So how should we judge Laurent Schwartz, the author of this book? He certainly is a revolutionary and challenges many of the cherished beliefs of the cancer establishment. I first met Dr. Schwartz at a meeting I chaired at Cambridge University in the winter of 2000. The meeting was entitled "Blue-sky cancer research". The meeting was funded by a group of merchant bankers and the invitees came from every walk of life. There was an inner circle of clinical and basic scientists and an outer circle of representatives from finance, economics, politics, philosophy and the arts. We all had two things in common: frustration with the pace of progress in cancer research and a willingness to think "outside the box". Many exciting ideas emerged from a long weekend of brainstorming and a road map for blue-sky research was proposed which not surprisingly failed to find financial backers from either the research councils or business. Perhaps some of the ideas were a bit wacky, but all those who attended and signed up to this portfolio of ideas were high achievers in their own fields, so as a group at least we weren't all mad and bad. I have described the outcome of this meeting in a paper published in "Prospect",[4] but I just want to quote some passages that adumbrate the message in this book.

All the rhetoric concerning cancer treatment and cancer research is couched in the terms of military conflict. The politicians trumpet the fact that we are winning the war against cancer. We talk about mobilizing resources and the cancer researchers with the highest public profile are described in terms normally reserved for military leaders. At the sharp end the poor patient never simply dies of the disease but loses the fight. This terminology fixes in the mind of the public and the politicians the idea that cancer is a foreign invader and therefore the war against cancer is aimed at destroying every last malignant cell.

---

[4] A new strategy for cancer, Baum M., Prospect, Februrary 2002, 44–48

This is a false analogy. The cancer cells are simply an undisciplined sub-stratum of our cells. Cancer is an inevitable component of the aging process and all of us at some time in our lives co-exist with latent cancer scattered around the body. No wonder the most aggressive modern treatments can end up killing the person, before killing the cancer. Medical oncologists often complain about the narrow therapeutic ratio within which they have to work – in other words the differences between the normal cell and the cancer cell are so small that they can rarely be exploited to the patient's benefit. No we are not winning the war against cancer and the aggregate mortality for malignant disease has barely changed since President Nixon declared war in 1971.

In 1971 President Nixon launched his war against cancer with the now notorious Cancer Act. In the same way that President Kennedy had promised to land a man on the moon in the 1960s, so Nixon would find a cure for cancer in the 1970s. I was working in Pittsburgh with Professor Bernard Fisher at the time and shared his deep scepticism about the possibility to deliver on this promise. When Kennedy promised the Americans that they would land a man on the moon we knew with a very high degree of accuracy where the target lay. All that remained was to create the technology based on the rocketry developed in the Second World War. By comparison, for cancer, we did not know in 1971 where the target was and the conventional non-specific cytotoxic therapies were equivalent to firing off rockets in random directions.

Undoubtedly there have been dramatic breakthroughs in the treatments of leukaemia, lymphoma and the childhood cancers. Thirty years ago all these would have been fatal but today we can expect between 50% and 75% cure rates (depending on sub-types). These followed on the classic experiments with animal models and cytotoxic chemotherapy in the 1960s and as it turns out this group of diseases is exquisitely sensitive to cytotoxic drugs. Unfortunately the very diseases that are most responsive to chemotherapy tend to be extremely rare and account for less than 5% of the total cancer burden.

Furthermore, the very success of chemotherapy in these rare cancers has paradoxically delayed progress in discovering effective treatments for the more common solid tumours such as colorectal and bronchogenic carcinoma, which have stubbornly refused to respond in significant numbers to these very toxic treatments. Instead of the reductio ad absurdum of persisting with high-dose chemotherapy to virtually lethal doses, we should try to learn from mistakes of the past and understand why rare cancers respond to chemotherapy but common cancers do not. All that aside, what else of importance can be learnt from trends in incidence and mortality? The incidence of some cancers falls or rises for reasons that are not understood. For example, at one extreme, stomach cancer, which is very common in Japan, is rapidly disappearing in the West. We now have a generation of young surgeons who have never cut their teeth on a radical gastrectomy for this horrid disease. There must be lifestyle reasons or environmental changes to explain what has happened but these are still a mystery. At the other extreme, malignant myeloma (a malignant process effecting the bone marrow) is increasing worldwide, almost as if there had been a viral vector.

In contrast, however, the trends in the incidence of lung cancer and lung cancer mortality are easy to explain. In Britain and the USA, lung cancer mortality rates for men fell rapidly once there was a clearly established link between smoking and the disease. Tragically, many young women are now taking up the habit, as a result of which deaths from lung cancer have overtaken deaths from breast cancer amongst women in North America and Britain. There is only one common cancer that can truly be viewed as a success story of modern treatment – breast cancer, where treatment has contributed to about a 30% reduction in mortality rates over the last 15 years. This can largely be ascribed to the fact that breast cancer, unlike the majority of other cancers, is peculiarly sensitive to the level of sex hormones in the circulation and we now have safe and effective drugs that can modify these oestrogen levels. But as far as the other common cancers are concerned – lung cancer, bowel cancer, ovarian cancer, cervical cancer, prostate cancer – little in the way of therapeutic advances has been seen in spite of the billions of dollars thrown into cancer research and treatment since Nixon's initiative of 30 years ago. Why this has been the case and how we might break out of the impasse are issues explored in this book.

Let us now consider the received wisdom about cancer that marks out the conceptual constraints which determine all programmes of fundamental research into the problem.

Cancer is a molecular problem. Either by inheritance or by exogenous factors that target us through life, specific cells accumulate sufficient mutations that have escaped natural DNA repair mechanisms to exhibit the properties of a malignant phenotype. The cancer cell then has the capacity for promiscuous growth, immortality, infiltration, dissemination and the establishment of remote colonies. The cancer then kills by unrestrained growth in a closed compartment, e.g. brain tumours, or by its metastases that destroy vital organs such as the liver, lung or bone marrow. The triumphalistic reception to the news that the human genome had been decoded, with President Bush and Prime Minister Tony Blair standing side by side in front of banks of cameras to announce the news, was fuelled in part by the assumption that the complete understanding and cure of cancer was just around the corner. Sadly, unlike the war in Iraq the war against cancer is still being lost. I do not wish to minimize the achievement of cracking the code; it is indeed a triumph of human ingenuity, technology and perseverance. Unfortunately there are even taller mountains to conquer before we can begin to understand the cancer cell and its relationship with the host's body. It is common parlance to talk about the tumour–host relationship, yet that is a giveaway, disclosing the assumption that the cancer cell is an alien parasite. In fact one of the themes of this book is the similarity of the cancer to its progenitors and the plasticity of the cancer cell with its capacity to redifferentiate to a normal cell or even another fully differentiated adult phenotype.

The genetic code merely reads off the instructions for lining up the amino acids in the appropriate sequence for a specific protein. This string of amino acids then has to fold into exotic shapes before the protein can function. The understanding of how the folds are determined and how the three-dimensional shape determines function is known as proteomics and this is the next great challenge. But even that is barely scratching at the surface of the awesome task of understanding how tens of thousands of these complex molecules interact with each other to create a cell and how millions of cells interact with each other to create you and me. Only then, when we have recreated the human subject from the rubble of molecular reductionism, will we have a glimmer of understanding of how the fault in the gene or the fault in the folding of a protein or the fault in the cross talk or geometry of cell-to-cell interaction leads to the human suffering of cancer. Imagine the task of rebuilding a functioning Boeing 747 from the scattered components of a deconstructed plane without a blueprint and a less than complete understanding of the physics of flight, amplify that challenge a million times and that gives you a feel of the magnitude of the task ahead. But then pause for a moment and consider that the genetic abnormalities of cancer may not be the direct cause of the malignant transformation, but in part an indirect consequence of faults at different levels in the hierarchical organization of the body just described, feeding back to destabilize the human genome. If that is the case then much of the current programme of cancer research is misdirected. It reminds me of the story of the drunk trying to find the key to his house under a lamppost some distance from his front door. When challenged on this he declared with hurt dignity that this was where there was sufficient light to see the key. The techniques of molecular biology certainly shine sufficient light but maybe the key to cancer is in the penumbra of that zone.

With all this in mind how should we approach this book? The first thing I ask is to crave your indulgence. I am often surprised by the hostility of the response when I politely challenge a speaker at a scientific conference on the assumptions that underpin his or her favourite hypothesis. We are sometimes wedded to our ideas with greater fidelity than we are wedded to our spouses.

Next, note the vast accumulation of observations described by Laurent Schwartz that cannot neatly be incorporated into the dominant conceptual model of cancer. As Thomas Huxley once stated, "The great tragedy of science is the slaying of a beautiful hypothesis by an ugly fact". Like Dr. Schwartz, I believe there are now so many outlying facts that we are currently at a point of crisis with the molecular paradigm of cancer. But simply wrecking a paradigm is not sufficient; we need a new set of beliefs that can explain the successes of the past yet incorporate all the outlying facts that spoil the picture. I think that the author is right in forcing us to look at epigenetic phenomena, the geometry and physics of cell polarity and physical constraints; he also make powerful arguments to reconsider Warburg's hypothesis concerning anaerobic metabolism and glycolysis. Like the author, I agree that we need to look at the new mathematics of nonlinear systems and consider how one of the stigmata of cancer is the loss of the perfect fractal geometry of the duct and vascular systems. Where I part company concerns the nature of latency. If we consider the breast in women, the prostate in men and the thyroid in both sexes, then all of us at most times carry latent (usually referred to as in situ) cancers.[5] They exist without evidence of inflammation and are known to demonstrate loss of heterozygosity and evidence of genetic mutations. However, only a minority of such lesions progress to invasive cancer, otherwise all of us would die of the disease. However, if we consider these as a necessary but not sufficient condition for cancer and then plug in the Schwartzian hypothesis of inflammation, anaerobic metabolism and cellular polarity, we have a beautiful model of carcinogenesis that can incorporate most of the observations described in this book. Furthermore, the hypothesis lends itself to refutation by the appropriate experiments or better still spectacular cor-

[5] What is the evidence that tumours are angiogenesis dependant? Folkman J. Journal National Cancer Inst. 1990, 82:4–6

roboration of what after all is a "bold conjecture". If the author were correct, then simple drugs that inhibit glycolysis would have a place in cancer therapy.

Let me finish this introduction with an anecdote. Last year a 52-year-old nurse came to see me for a second opinion. Nine months earlier she accepted an invitation for mammographic screening. The X-rays showed a small focus of microcalcification, which was biopsied and shown to be duct carcinoma in situ. She was advised to have surgical excision but refused. In the short time between that biopsy and her visit to my clinic the breast rapidly filled with invasive cancer. This was not the "natural history" of the disease, but to my way of thinking a latent pathology with molecular damage, which left to nature might never progress but once perturbed by the inflammatory response to the trauma of a biopsy it releases its malignant potential via the Schwartzian pathways.

Professor Michael Baum MD; ChM; FRCS; FRCR
Emeritus Professor of Surgery
and Visiting Professor of Medical Humanities,
University College London

Acknowledgements

This book is the result of extensive work done by scientists, most of them working at the Ecole Polytechnique near Paris.

Vincent Fleury is a young but established physicist specialized in morphogenesis at the Laboratoire de la Physique des Milieux Condensés. In his last book, *Des pieds et des mains* (Flammarion, Paris, 2003), he suggested the importance of physical forces in regulating cell growth, division, and differentiation.

Habibou Maitournam, from the Laboratoire de Mécanique des Solides, and Xavier Wertz confirmed that bone growth and differentiation is regulated by physical constraints.

Marcel Filoche works in the same department as Vincent Fleury, at the Ecole Polytechnique. He also specializes in fractal growth. He has done most of the research on the epidemiology of cancer. It is his idea that the aging individual is a system near "critical state." His results were later confirmed by two other mathematicians, Mireille Summa and Francoise Goupil, both from the University of Paris Dauphine.

Jean Marc Steyeart from the Laboratoire D'informatique of the Ecole Polytechnique, John Saul, a geologist and Maurice Israël, a recently retired biochemist as well as Ted Sanders have contributed a great deal to the formalization of most hypotheses. I also wish to thank Antonello De Martinello and Bernard Drevillon for their constant help.

Jacques Leibowitch played a key role, first in the discovery of the AIDS virus then in the partially successful treatment of that infection. He must be credited with the idea that inflammation and transient glycolysis are likely to be synonymous. Furthermore, he guided me in the subtleties of the molecular biology of cell division and into the partially forgotten work of Otto Warburg.

Vianney Pichereau, a bacteriologist from Caen, has been of great help in understanding the importance of catabolic repression both in bacteria and in human cells.

Most clinical discussions were held by Jacques Ballosso, Bernard Brun, Philippe Icard, François Baillet, Xavier Sastre, or Sylvie Delanian.

The experimental confirmation of the antitumor effect of 2-deoxy-D-glucose has been generously established by Philippe Genne at Oncodesign, a private company in Dijon.

Contents

Introduction

CHAPTER 1

This book is intended for cancer specialists, clinicians, and researchers. Its aim is to offer a new paradigm, a new cancer equation, not in order to create controversy, but simply in the service of science. I address myself to a community of scholars, although I am not a scholar. I am a field worker and my field is cancer.

What I do is treat patients whose chances for survival are very slim. Theory is not what makes up my daily practice. Instead, I deal with pain and fear but also with hope and success. For I feel almost completely helpless in the face of this disease. A few exceptions aside, we know neither the cause of cancer nor even the most basic elements of what would constitute effective treatment. Therefore, I decided to explore these questions by taking a path less traveled by cancer specialists: that of mathematics, physics, and chemistry. Oncology has not yet ventured into these areas.

My role as the only practicing oncologist in this group was simply to ask questions, the answers to which make up this book.

Mathematicians have been heroes in our culture. In the 1930s, a group of young graduates of the Ecole Normale, with only their enthusiasm and friendship to sustain them, created the mathematical equations of the modern age. They did it outside of the institution, but not to contest the institution. Their group was baptized Bourbaki. Thanks to Anne Fagot and Jean-Paul Amann and their gracious welcome, our Bourbaki is the "Collège de France."

Our work was carried out under no particular banner, without regard for the politically correct. We aimed neither to praise nor to criticize anyone. We were simply looking for the truth. None of us is a real expert in cancer research; therefore, the text may contain some errors and inaccuracies. As in psychoanalysis, we had to explore certain facts which are obvious but are drowned in the overabundance of published material (over twenty million citations on the Internet and one million scientific articles on Medline!). Like the raiders of the 1970s, who thought it best to buy an oil company at a discount rather than carry out risky and expensive operations, we decided to use the scientific data banks at our disposal. Thus, our work offers few encouraging scientific experiments or promising new drugs, but is rather an attempt to understand the existing data. We feel that the facts have already been demonstrated, but that their importance has not been realized. Moreover, proof, like the truth, tends to be forgotten. We admit that our method diverges from the classical. Usually, painstaking experimentation is used to confirm an intuition. But it is also possible to develop a theory and compare it as objectively as possible to the existing data. For instance, Copernicus never carried out any measurements or experiments, but he analyzed brilliantly the meticulous, methodical measurements recorded by Tycho Brahe.

To date, cancer has been seen as pathology of the cell. We are still under the spell of the tremendous success of antibiotics and vaccination. The enemy is a cell: either foreign (a bacterium) or personal (cancer).

Subsequently, all research has focused on one enemy: the cancer cell seen as a sort of particularly malignant bacterium. Such cells are scrutinized from every angle to try detect their cancer-causing properties. Treatment consists of targeting and killing them: The abnormal cell is subjected to anticancer chemotherapy (named after antitubercular chemotherapy).

Certainly, identifying cancer as a cellular disease was effective for a period of time. Some cancers could be made to regress by destroying cells, and manipulating cells could also provoke new cancers. It seems obvious that the sun revolves around the earth: It rises every morning in the east and sets every evening in the west. But tremendous intellectual contortions and a complete reversal of common sense perspectives were needed to explain the revolution of planets around the earth.

Similarly, the cancer cell could be innocent after all. When we inject a white mouse embryo with a cancer cell from a black mouse, a healthy black-and-white mouse will be born. If we inject cancer cells into the veins of the tails of young mice, these cells take hold in the first available filter: the liver. There is no cancer, only cancer cells transformed into normal cells living harmoniously in the liver of a healthy animal. All these complicated maneuvers (reversion genes, idiopathic regression, and homeostasis) to explain that something is wrong with the model! Worse yet, these type of data rediscovered here and there were presumed to be false since they made no sense.

The thread we followed on our journey consisted of understanding the hidden link between each definition of cancer (star-shaped tumoral structure, uncontrolled proliferation, loss of cellular differentiation). For a cellular biologist, cancer is loss of contact inhibition, for a molecular biologist, cancer is an oncogene and tumor suppressor gene anomaly, for an epidemiologist, cancer is the primary cause of death entered on a death certificate. A sociologist or a psychiatrist will have yet another definition.

Are these definitions referring to the same phenomenon? The pointillist painters understood that the play of form and color is clearer from a distance, that the whole picture is better seen when you have perspective.

Our perspective was based on the one most relevant to a physician: Galen's crab. Galen was the first to formulate a definition of cancer: a tumor which strangles the gland with its multiple extensions. The idea of cancer as an abnormality similar in form to a crab is already present in this definition. The clinician suspects cancer on the basis of physical characteristics. Like a crab, cancer is a hard, infiltrating mass. But its definition also specifies that cancer is inextricable from the gland in which it originates. For example, in order to remove a benign prostate tumor, the doctor inserts his finger between the tumor and the healthy prostate. A benign tumor is easily detached. But a surgeon cannot accurately delimit cancer because there is no clear separation between healthy and diseased tissue. There are cancers of various types: some are very infiltrating with no palpable tumors, others are pulpier. But all of them present the same dendritic extensions inextricable from healthy tissue.

Proof of cancer is provided by the anatomopathologist. The diagnosis is based on a microscopic examination. For him, cancer is a disorganization, with cellular layers infiltrating healthy tissue sometimes massively and sometimes more diffusely. The pathologist also looks for Galen's crab, albeit on a different scale.

The epithelium, which lines the mucosa in which the vast majority of cancers originate, resembles the pavement of the old streets of Paris: cobblestone against cobblestone. Like these stones, cells are not independent: they cannot escape from their neighbors, to whom they are tied through a system of anchors, the integrins. Epithelial cells can only divide when there is deficiency, a dead cell, or a wound. If the epithelium grows faster to compensate losses, as in embryogenesis, it can only grow by becoming elongated or bent.

Cancer is a barricade: stone upon stone, forming a pile. Knowledge of physics suggests that the very particular form of cancerous growths could be attributed to a single factor. In the barricades, the stones pile up; in cancer, cells climb on top of each other. The anchorage systems we mentioned earlier are altered. Simply put, the form seen in cancer could be explained by the repeated passing of cells over the cellular layers in which they originated. It is simple and it might be all there is to it.

Let us look at the proof. The anchorage systems between cells, which prevent them from changing levels, have been well known for some time. Many molecules inhibit or selectively destroy these anchorage systems, allowing the cells to change levels; these molecules are cancer producing.

Cellular biologists most often define cancer as loss of contact inhibition. Normal cells multiply in a Petri dish; when they reach the edges of the dish, they stop growing. In contrast, cancer cells are not inhibited by contact with the edges of the dish; they continue to divide, change layers, and pile up on top of each other. This is loss of cellular polarity. Galen's crab is a direct consequence of this loss of cell polarity.

But does this loss of cellular polarity suffice to explain unlimited cancer growth?

Physics experts tell us that a normal cell only divides if it has room to do so; pressure from its neighbors is too great. But if the cell next to it dies, the pressure lessens and the cell divides. When a cut causes a wound on the skin, surface pressure is diminished and epithelial cells divide until the edges of the wound come together and pressure reverts to normal.

Orthopedists provide another proof of the capital importance of physical constraint in the regulation of cell division. To lengthen a bone, they use traction. To stop it from growing, they use a cast. Who does not remember the bound feet of Chinese noblewomen before the Maoist revolution? To show their social position, young girls wore shoes that were too tight and prevented their feet from growing. Since these girls could not move about easily, they had to be served.

An organ grows in volume relative to the hard envelope in which it is contained. On the sixth floor of the Paris faculty of medicine there is a museum of horrors where aborted fetuses with various anomalies are conserved. A skull that did not close in time produced a disproportionate brain protruding outside the head! A diaphragm that provides no barrier allows a disproportionate lung to invade the abdomen!

Medical literature has occasionally published texts showing that physics governs cell division. When cells are under sufficient pressure, they are prevented from multiplying; by contrast, stretching causes them to divide. These are difficult manipulations, not well thought of in contemporary scientific circles.

It is difficult to study this type of constraint, since this work draws on a culture other than that of the biologist. The biologist assumes that growth factors or growth inhibition factors control all cell division. These exist in infinite numbers; their effects are often contradictory from one cell line to another or from one culture mode to another. This unusual inventory of elements seems endless. Today, this type of biology produces no obvious principles. For example, why is it that in adults one single cell replaces one dead cell? How can we explain this cellular homeostasis? Having reached adulthood, neither our skin nor our intestine, both of which undergo constant renewal, lengthens or shrinks.

This same constraint governs the growth of tumor cells, just as it does that of normal cells. Every clinician knows that cancer invades soft tissue more easily than hard tissue such as a bone, tendon, or fascia. However, contrary to normal cells, tumor cells can invade areas of lesser constraint. The normal cell is prevented from multiplying by the constraint exerted by its neighbors. But the tumor cell escapes this constraint by changing plane of division, and can continue to divide. It is able to escape by finding areas of lesser resistance. This appears to confirm that it is indeed loss of cellular polarity which accounts for tumoral growth.

Modern biology focuses on the discovery of a biological motor. But there is no need to understand the subtleties of the internal combustion engine to explain Sunday night traffic jams! Similarly, knowledge of the subtleties of the cellular "engine" is no doubt superfluous for understanding the growth of cancer.

But there is more to cancer than unrestrained growth. The cytologist who examines cell smears to detect a neoplasm defines cancer on the basis of loss of cellular differentiation. The tumor cell is different, often larger, with a shapeless nucleus. Cancer cells are said to be undifferentiated, meaning that they have lost their original programming. Breast cancer produces no milk; cancer of the salivary glands produces no saliva. This is likely to be an additional consequence of a change in spatial configuration.

A doctor sees countless examples of the importance of the extracellular constraints in cell differentiation. I am writing this text while vacationing in Brittany, a region where congenital hip luxation is common. The femur is dislocated upward in the pelvis, above its normal position. The result is an abnormal femoral socket, located too high in relation to the femur. This out-of-place articulation is fragile. It is as if the faulty position of the femur creates this articulation through repeated friction. Here is another example: following an accident, a portion of bone has disappeared. If there is no fixation surgery, the muscles on the remaining bone will confer contradictory movement to the proximal portion of the distal. A new articulation will form between the remaining bone portions. This pseudoarthrosis is usually painful and debilitating. Other examples: a silicone breast implant is quickly surrounded by synovial cells; the hands of a manual laborer are rapidly covered with calluses.

Development of the embryo and the fetus also depends on the environment. At birth, the epiphyses are made of cartilage. The progression of bone formation is an exact reflection of the child's age. Thus, when the police want to establish a young offender's age, an X-ray of the hand is taken. Mechanics experts tell us that transformation of one tissue into another, for example cartilage into bone, takes place based on local constraints. This intuition is confirmed by laboratory experiments: if you compress chicken cartilage, it will become ossified.

This whole process is antithetical to definitive genetic programming. In order to avoid saying that the environment controls the cellular mechanism, the euphemistic term "cellular plasticity" is currently in vogue.

Let us provide one final proof, of the type mathematicians call "ad absurdum." Cells in a Petri dish are well nourished, subjected to little mechanical constraint, and show almost no differentiation. In order to restore some differentiation, the cells must be bound by means of integrins: in other words, they must be constrained. One of the noteworthy failures of molecular biology resides in its inability to produce saliva, milk, or pancreatic juice in the laboratory.

In order to understand how the cell functions, we have engaged in learned investigations of the genome. Apparently, it is all a question of physics and biochemistry. But no matter how complex a genome may be, it is still subject to the laws of nature. Cell differentiation seems to be mostly a function of the cell's environment.

Let us return to clinical practice. One of the first questions a cancer patient asks is: "Why me?"

There again the role of cell environment is crucial. What the clinician sees is that cancer never develops in healthy tissue. Cirrhosis, alcohol related or not, prepares for liver cancer; chronic bronchitis paves the way for lung cancer; chronic parasitic infection precedes bladder cancer; chronic pancreatitis is at the origin of cancer of the pancreas. The more severe the chronic bronchitis, cirrhosis, or pancreatitis, the greater the risk of cancer. We all know that tobacco causes architectural changes such as emphysema, chronic bronchitis, and subsequent lung cancer. Even in nonsmokers, architectural changes such as tuberculous cavity, pulmonary fibrosis, and nickel- or chromium-induced fibrosis precede lung cancer.

I have never diagnosed cancer in a healthy tissue. As a young resident, I had to attend the autopsies of my patients. The lung of a bronchitis patient, just like the liver of a cirrhosis sufferer, was hard and wrinkled like old skin. Microscopic analysis confirms the thickening of the conjunctive tissue. Cirrhosis provides the best example. The epithelium is distorted; the tubular epithelium becomes nodular in the regeneration nodules. The greater the distortion of the epithelium, the greater the risk of cancer. The forces acting on the cells are no longer the same. In some places, the cells are more compressed than in others. In a zone of lesser constraint, the cell will divide and will escape from its neighbor by sliding away. Thus, the cancerous clone is born.

Patients with genetic developmental anomalies are more susceptible to cancer. Children with trisomy develop leukemia and solid tumors. What do chronic bronchitis, cirrhosis, bilharziasis, and trisomy have in common? They all present a thickening of the conjunctive tissue underlying the epithelium. The ultrasound performed at 3 months' gestation is intended to detect trisomy. The doctor measures the increased thickness of the fetal nucha, that is, of the subcutaneous conjunctive tissue.

We have known for over 50 years that changing the architecture of a tissue is sufficient to induce cancer. For example, subcutaneous injections of cellulose, chemically inert but sharp, are cancer producing. The same cellulose is harmless when the borders are rounded. Asbestos, a fireproof material, comes from different countries. Asbestos fibers from some countries are less sharp than those from other countries. Sharp fibers are much more dangerous than blunt fibers. This difference in origin, and consequently in form and in harmfulness, has been the main defense argument of the asbestos lobby. For the same reason, nuns no longer wear "horns" as a headdress: the constant friction caused ear cancers. Man's best friend, the dog, provides another example. Veterinarians insert small radio transmitters in the dog's ear so that, if lost, he can be found. The form of this transmitter determines whether or not the animal will develop cancer.

Physics is only one aspect of the picture. Most cancers are not caused by foreign bodies. Here again, history provides insightful clues as to why loss of cell polarity appears to play such a key role in cancer.

At the turn of the nineteenth century, the Prussian state coordinated all efforts in view of applying the results of fundamental research not only to industry, but also to the preparation of a war with France, seen as inevitable. In today's jargon, we would describe such a project as "fundamental research at the service of the military–industrial complex."

These efforts produced the first synthetic dyes for the textile industry (German farmers were already raising sheep and planting hemp and linen), the first nitrous fertilizers for agriculture (German soil is often poor), as well as the first synthetic drugs such as aspirin. The Germans also developed a chemical industry, which is still the most powerful in the world.

They took an interest in the chemistry of the living organism. The vision of that era was simple: in order for a cell to live, it needs energy. This is what led to the discovery of the importance of cellular breathing and of oxygen metabolism. It was this gigantic project that led to the discovery of enzymes, which convert sugar, proteins, and fat into energy. Any medical student can attest to the difficulty of remembering these cycles which fit together like Russian dolls, and which are still known by the names of the German chemists of that period.

And what could be easier, when you understand life, than to destroy it? From the start of the First World War, these scientists lent their knowledge to the German war effort, with the resulting devastation caused by yperite and other combat gases.

As a medical student in the late 1970s, I had occasion to see the terror expressed by patients at the mere mention of gases, and to see the pulmonary sequels of these gases 60 years later.

Yet another poison was synthesized: Zyklon B, used so "successfully" in the death camps of the Second World War. Needless to say, this research on cellular breathing mechanisms – whatever its original merits – suffered greatly from the ensuing madness.

Let us come back to cancer. Otto Warburg measured oxygen in cancers. As early as 1920, he knew how to inject tumoral suspensions into the peritoneums of mice and how to measure their gas concentration. He understood that cancer is a disease of cellular breathing. He also understood that all cancer-producing substances (arsenic, tars, and cyanide) cause hypoxia. Either oxygen cannot reach the cell or it cannot be utilized. In the 1920s, Warburg had identified these two phases: first, hypoxia alters cell metabolism; second, if the cell survives these anomalies, the latter will produce cancer. Despite the Nobel prize awarded to Otto Warburg, his work has largely been forgotten. Almost a century after his first discoveries, cancers are still hypoxic.

What came first, the chicken or the egg? Today, hypoxia is commonly thought to be a consequence of rapid tumor growth exhausting the vascular supply. This could be wrong. On the contrary, cancer could be a logical consequence of hypoxia. How can we be sure? It is notoriously difficult to measure cellular concentration in oxygen and is even more difficult to modulate it.

My perspective is that of the clinician. We know that inflammation and consequent fibrosis and tissue disruption pave the way for cancer. It may well be that inflammation and transient hypoxia are synonymous.

The definition of inflammation has not changed since Galen (for close to 1,800 years). It is still defined as pain, redness, and heat in tissues. But Galen had only his hands for instruments. He knew nothing about chemistry and oxygen.

Just like cancer, inflammation has numerous causes, which, at first glance, seem very different. What do an infection, a foreign object such as a splinter, irritation caused by a chemical, and an allergen such as tiny dust mites living in bedclothes have in common?

A microscopic examination reveals that the tissue is invaded by white blood cells, that dilated blood vessels are responsible for redness of the skin, and that fibroblasts secrete collagen.

Critical analysis of extensive data banks allows us to conclude that oxygen reduction alone may be sufficient to explain white blood cell invasion, collagen deposits, and dilated vessels.

Our hypothesis is that Galen's inflammation is nothing more than temporary hypoxia. Here again, we may be rediscovering the wheel. No need to carry out experiments, we already have the proof. In cases of heart attack, peripheral tissues located far from the heart, such as skin, internal organs, and muscles, present inflammation. Once again, there is lymphocyte proliferation, collagen deposits render the viscera fibrous, and blood vessels are dilated. In animals, an artificial heart can correct this hypoxia: the inflammation simply disappears.

Hypoxia results in extracellular matrix deposition. Fibrosis, the ubiquitous complication of untreated inflammation, perpetuates hypoxia and paves the way for cancer.

The effect of oxygen on bacteria has been known for a long time. A change in milieu can trigger cell multiplication. In Petri dishes, animal cells can be exposed to air poor in oxygen. Human cells are full-fledged organisms, which behave according to the chemical and physical milieu in which they live. When the concentration of oxygen is reduced, cells undergo morphological changes (loss of cell differentiation) and proliferate. This deprivation causes a decrease in cell integrin and cell-to-cell contact. The epithelial cell may loose its polarity. This is the bridge between biology and physics.

Cancer is commonly thought to result from multiple anomalies of the genome. A carcinogen damages a few essential genes and cancer develops. Cancer as a consequence of genetic anomalies has produced marketable notions: cancer genes, tumor suppressor genes, and the necrosis factor of cancer previously known by the less poetic name "cachexia factor." Everything seems to point to the essential role of genes in cancer. First of all, hereditary cancers. The genes responsible have gradually been cloned. Counseling and prenatal diagnosis are already a reality. At the same time, it has been observed that cancer cell genomes present many anomalies: mutations, suppressions, and genetic amplifications. The assumption is that a few mutations can cause a cell to become cancerous.

Attributing cancer to genome anomalies, as is done much too often, is a mistake in most cases. First, most cancers are not hereditary. For example, less than 5% of breast cancers are of genetic origin. Activation of oncogenes is not specific to cancer. It is frequent in embryogenesis, and even in adult life. One study has found oncogene activation in 70% of normal breasts.

Forty years ago, in the early stages of molecular biology, cloning a gene was a gigantic enterprise. Mutations in cancer were rare. Today the process has been simplified. We speak of 100,000 mutations per cancer cell. Far too many.

Our work consisted of changing perspective, forgetting the genome of the cell in order to concentrate on the cellular environment, changing the focus from cancer as a cellular anomaly to cancer as the disease of an organ. A new perspective is valuable only if it conforms to reality. Changing perspective does not negate the existence of genomic alterations; it includes them in a larger picture. In the sixteenth century, the inconsistencies of an earth-centered cosmology were obvious to all the experts. But, the Copernican revolution was finally accepted only because it confirmed what anyone could observe directly: that the sun rises in the east, that there are about 28 days in a lunar month...

Can a modification of the cell environment disturb the genome? Proof of this already exists. If you deprive a bacterium of food, it will generate spores. If you heat, chill, suffocate, or starve a cell, if you stretch it or compress it, mutations and a reorganization of the genome will result. All these manipulations are designated by the term "cellular stress." Whatever the exact nature of the stress, it activates the heat shock proteins responsible for genome instability. Chronic friction caused by renal lithiasis – although chemically inert – is enough to produce multiple mutations of the integument epithelium initially, and then cancer. But mutations are just as frequent with these cancers as with any other neoplasm.

This deprivation in oxygen also results in mutation, oncogene activation, and inhibition of tumor suppressor genes. The activation of these defense mechanisms is, for a time, not the cause, but the consequence of cancer. Subsequently, this oncogene activation will, in turn, increase cellular hypoxia and accelerate carcinogenesis.

In the following chapters, I will prove that cancer is a direct consequence of changes in constraint. These changes are a consequence of hypoxia. As I am trying to convince a learned audience, I will be using more technical terms and referenced sentences.

Cancer is both a fascinating conundrum and a merciless killer. Our only real goal is not to solve a riddle but to contribute to a solution. In the final chapters, we will see how this tentative redrawing of the picture can lead to effective treatment.

Are Cancer Cells Malignant Per Se?

CHAPTER 2

Our reasoning is shaped by the great successes of the last century. Just like a microbe causes infection, cancer is supposed to be the consequence of genetic injuries inside a human cell. But cancer cells do not summarize cancer.
There may be a persistent tumor mass following successful chemotherapy. These cancer cells have changed destiny and stop dividing. Similarly, cancer cells injected into a healthy animal or an embryo often differentiate into normal cells. Cancer cannot be ascertained by cytological examination alone.

Today, it is universally accepted that cancer is a disease of the cell. This was not always the case. Up until the middle of the twentieth century, cancer was considered a disease of the tissues. Hence, treatment consisted in the excision of the entire affected organ, as illustrated by the radical mastectomies advocated by Halsted. Language still preserves traces of the time when cancer was a disease of the organ. We still speak of cancerous tissue and cancerous tumor transplants.

Medical and biologic ideologies advance hand-in-hand. Alexis Carrel (1) focused his research on the organ rather than the cell, and in 1938 he wrote: "For the first time, medical science is capable of apprehending bodily structures in the fullness of their reality, of understanding how the organs form the organism, and how the organism grows, ages, heals its wounds, resists disease, and adapts itself with marvelous ease to changing environment." To say that cancer results from extracellular anomalies and not from intracellular lesions is to revive old ideas and, at the same time, introduce a new state of affairs.

Today, the obvious claim is that cancer originates in irreversible lesions in the DNA. The reason for this change in paradigm is not based on any proof, but rather on progress in cell and molecular biology.

We are now able to extract human cells and keep them in culture in a Petri dish for dozens of years. Everything is done to improve efficiency. But just as in traditional agriculture, what comparison is there between transgenic corn and the original plant grown on the high Mexican plateau? We know how to cultivate cells, but not organs. Therefore, the cell and its mutations explain carcinogenesis, chemosensitivity, and resistance to treatment. It is likely that oncology tenets would be different if we knew how to cultivate organs.

Cellular Theory and the Influence of Infectious Diseases

As early as 1946, Hammond (2) was able to sustain and eventually to grow mouse embryo cells. These cells, laid out in a Petri dish, multiplied in a serum-enriched milieu. Soon, the procedure was repeated with tumor cells. In 1951, in Baltimore, a young lady, Henrietta Lacks, was diagnosed with cervical cancer. Tumor cells were surgically removed and cultured in vitro. The cells were indeed malignant and Henrietta Laks died 8 months later. But Henrietta Laks's cells (HeLa cells) are still alive and well in laboratories all over the world nearly 50 years after her death.

Normal adult cells do not lend themselves well to cellular culture, and can only be cultured for a limited time. The immortality of tumoral cells seems to constitute the very nature of cancer.

In an era primarily defined by progress in the fight against infection, the same scientists grow bacteria and tumor cells. Gradually, a change of paradigm took place: cancer became a disease of the cell. The cancer cell replaced the bacterium. The term "antitumor antibiotics" is a reminder of this shift.

Molecular anticancer activity is tested on cancer cells like HeLa. Anticancer activity is defined as the ability of molecules to kill cancer cells first grown on Petri dishes ("in vitro") then injected in animals ("in vivo"). The hypothesis of the cellular nature of cancer becomes dogma.

This assumption has been reinforced by the fact that cancer can be transmitted from one animal to the other by the simple injection of tumor cells. But are we transplanting isolated cells or a tissue?

A single cancer cell injected into an animal does not develop into cancer. Like in a transplant, thousands of cells are required for tumor growth (3–6). Single cells injected in the blood stream regroup after injection (7, 8) to form a cancerous clot. Treatments, which simply inhibit cell-to-cell attachment, drastically decrease the ability to form tumors (9, 10). Just as for the graft of an organ, tumor cells only grow in syngeneic animals or in an immunosuppressed host (3, 4).

Tumor Cells Alone Are Not Enough for Cancer Diagnosis

The diagnosis of infection is based on finding of a germ. On the contrary, the finding of a single "cancer cell" is not sufficient for a diagnosis of cancer.

Today, the pathologist establishes a diagnosis of cancer through microscopic examination of a tumoral fragment. This fragment is usually obtained after surgical biopsy or from partial or total resection of a suspicious area.

Under the microscope, cytologic and architectural anomalies coexist. Pathologists have described cellular anomalies such as an increased number of mitoses, the presence of nucleoli in the nucleus, or an increased nucleus–cytoplasm ratio.

These cellular anomalies are not specific to cancer (11). To diagnose cancer based solely on the observation of cytologic anomalies is potentially dangerous, given the high risk of error. Inflammation can be responsible for the same cytologic anomalies as an aggressive form of cancer. This is why, when a cervical smear is suspect, the patient is treated with antibiotics. If the cytologic anomalies persist, only a biopsy can establish malignancy. The biopsy will confirm the cytologic anomalies and, above all, will reveal cancerous tissue structure.

Neither can a biological marker secreted by a cancer cell establish cancer with certainty. This explains the well-known unreliability of cancer detection based on blood level tests. Just as there is no cell that can confirm cancer by its shape alone, so there is no protein whose presence alone indicates cancer. Of course, some proteins (e.g., ACE, alpha-fetoprotein, beta-HCG) are secreted in large quantities by metastatic cancers. But these proteins are also synthesized by normal cells, which explains specificity and the failure of cancer screening through blood level tests. For example, 70% of apparently healthy rats express a mutation of the H-*ras* gene in the normal mammary epithelium (12). Prostate-specific antigen (PSA) proves the rule, it is a poor screening test.

The Often Normal Life of Cancer Cells

Multiple clinical observations suggest that cancer cannot be summed up as an irreversible disease of the cell. Teratocarcinomas are highly malignant tumors arising from poorly differentiated embryonic cells. Illmensee (13) demonstrated that a malignant teratocarcinoma cell could generate normal tissues and organs when transplanted into early embryos. The blastocysts developed into viable mice that were mosaics of normal and teratocarcinoma cells. Rats injected subcutaneously with rat hepatocarcinoma cells developed tumors; however, when injected in the liver, similar hepatocarcinoma cells were integrated into the normal liver parenchyma (14). The reason for this cell adaptation, whether genetic reprogramming or a change in the physical microenvironment, is unknown.

McCullough (15) injected transformed rat epithelial cells in the liver parenchyma. The tumorigenic potential was dependent on the age of the host. It appears that aging of the liver microenvironment was necessary for tumor growth (15). Tumor cells injected into the liver of young rats differentiated into normal-looking benign hepatocytes.

These experimental data are confirmed by clinical practice. Successful chemotherapy of human germ cell cancer and neuroblastoma is occasionally associated with residual tumors that on biopsy reveal only differentiated cells surrounded by fibrosis (15–17).

Similarly, the literature has described hundreds of cases of spontaneous remission; for no apparent reason, measurable cancers confirmed at histological examinations, of varying histology types (but most often kidney cancers and melanomas), regress or disappear completely. Explanations for this varied from an efficient immune system to a fighting spirit, to divine intervention.

Normal Cells May Be Immortal Too

One of the foundations of cellular cancer theory has been the immortality of tumor cells. The normal cell is condemned to an inevitable death, while the tumor cell appears immortal. This theory is based on an obvious contrast: endless multiplication of tumoral lines, and the difficulty of growing normal cells in vitro. Fifty years after Henrietta Laks's death, cancer cells are still multiplying, while the corresponding normal cells have long since disappeared.

In reality, cell immortality does not explain cancer. Tumor cells are mortal. The presence of necrosis is often associated with malignancy. The more aggressive the tumor, such as a glioblastoma, the greater the necrotic intratumoral component.

Apparent immortality is no longer the prerogative of malignancy. Since the first tumoral lines were assembled, it has become a standard assumption of biology that there is an intrinsic fixed limit to the number of divisions that normal human cells can undergo before they senesce, and this limit is in some way related to aging of the organism (18). Fibroblasts, like other cell types, age in vitro and display a limited potential for cell division in culture. For example, mesenchymal cells from the dermis of an embryo have a doubling potential of 50–60 times in vitro (19).

The in vitro evidence is countered by estimates that the number of cell divisions in a normal human being are several orders of magnitude higher than the in vitro limit, with no indication of the degenerative changes seen in culture. Serial transplantation experiments in animals also exhibit many more cell divisions than the in vitro studies, with some indicating an indefinite replicative life span (18).

Furthermore, not all cells experience this fixed limit. Epithelial cells are grown in culture, in vitro, in infinite quantities, for skin strips needed to cover serious burns. Embryonic or adult stem cells, just like neoplastic cells, can multiply endlessly.

In fact, this limited life span in vitro may be an artifact. Cells are severely stressed by enzymatic dispersion and sustain cumulative damage during serial subcultivations. The evidence includes large increases in cell size and cell heterogeneity, reductions in replicative efficiency at low seeding densities, appearance of abnormal structures in the cytoplasm, changes in metabolism, and DNA changes not seen in vivo. Rubin (18) proposes that the limit on the replicative life span is an artifact that reflects the failure of diploid cells to adapt to the trauma of dissociation and the radically different foreign environment of cell culture.

References

1. Lindbergh C, Carrel A (1938) The Culture of Organs. In: Hoeber P (ed) New-York, USA
2. Hammond J (1949) Culture of mouse embryos using an egg-saline medium. Nature 163:28–32
3. Price JE (1996) Metastasis from human breast cancer cell lines. Breast Cancer Res Treat 39:93–102
4. Murphy P, Alexander P, Senior PV, Fleming J, Kirkham N, Taylor I (1988) Mechanisms of organ selective tumour growth by blood borne cancer cells. Br J Cancer 57:19–31
5. Tominaga T, Yoshida Y, Kitamura M, Kosaki G (1987) Liver metastasis of colon: 26 cells implanted into the superior mesenteric vein in mice. Jpn J Cancer Res 78:846–850
6. Kikkawa H, Imafuku H, Tsukada H, Oku N (2000) Possible role of immune surveillance at the initial phase of metastasis produced by B16BL6 melanoma cells. FEBS Lett 467:211–216
7. Fidler IJ (1973) The relationship of embolic homogeneity, number, size and viability to the incidence of experimental metastasis. Eur J Cancer 9:223–227
8. Lione A, Bosmann HB (1978) Quantitative relationship between volume of tumour cell units and their intravascular survival. Br J Cancer 37:248–253
9. Meromsky L, Lotan R, Raz A (1986) Implications of endogenous tumor cell surface lectins as mediators of cellular interactions and lung colonization. Cancer Res 46:5270–5275
10. Glinsky GV, Price JE, Glinsky VV, Mossine VV, Kiriakova G, Metcalf JB (1996) Inhibition of human breast cancer metastasis in nude mice by synthetic glycoamines. Cancer Res 56:5319–5324
11. McGoogan E, Colgan TJ, Ramzy I, Cochand-Priollet B, Davey DD, Grohs HK, Gurley AM, Husain OA, Hutchinson ML, Knesel EA Jr, Linder J, Mango LJ, Mitchell H, Peebles A, Reith A, Robinowitz M, Sauer T, Shida S, Solomon D, Topalidis T, Wilbur DC,Yamauchi K (1998) Cell preparation methods and criteria for sample adequacy. International Academy of Cytology Task Force Summary. Diagnostic Cytology Towards the 21st Century: An International Expert Conference and Tutorial. Acta Cytol 42:25–32
12. Cha RS, Thilly WG, Zarbl H (1994) N-Nitroso-N-methylurea-induced rat mammary tumors arise from cells with preexisting Hras1 gene mutation. Proc Natl Acad Sci USA 91:3749–3753
13. Illmensee K, Mintz B (1976) Totipotency and normal differentiation of single teratocarcinoma cell cloned by injection into blastocysts. Proc Natl Acad Sci USA 73:549–553
14. Coleman W, Wennerberg AE, Smith GJ, Grisham JW (1997) Regulation of the differentiation of diploid and aneuploid rat liver epithelial cells by the liver microenvironment. Am J Pathol 142:1373–1382

15. Mc Cullough KD, Coleman WB, Smith GJ, Grisham JW (1994) Age-dependant regulation of the tumorogenic potential of neoplastically transformed rat liver epithelial cells by the liver microenvironment. Cancer research 54: 3668–3671
16. Hain SF, O'Doherty MJ, Timothy AR, Leslie MD, Harper PG, Huddart (2000) Fluorodeoxyglucose positron emission tomography in the evaluation of germ cell tumours at relapse. Br J Cancer 83:863–869
17. Miyauchi J, Matsuoka K, Oka T, Kamii Y, Honna T,Bessho F, Sasaki S, Melanowska J, Tsuchida Y (1997) Histopathologic findings of advanced neuroblastoma after intensive induction chemotherapy. J Pediatr Surg 32:1620–1625
18. Rubin H (1997) Cell aging *in vivo* and *in vitro*. Mech Aging Dev 98:1–35
19. Hayflick L,Moorhead PS (1961) The serial cultivation of human diploid cell strains. Exp Cell Res 25:585–592

Cell Growth: A Balance Between Glycolysis and Physical Constraint

CHAPTER 3

Human cells are exposed to biochemical and physical constraints. In the presence of oxygen, glucose is degraded into energy, water, and carbonic gas. At lower concentration of oxygen, by contrast, glucose is incompletely degraded and metabolites are produced. Some metabolites of glucose react to form amino acids, lipids and nucleic acids. These amino acids in turn form proteins, the lipids are transformed into hormones and the nucleic acid provides DNA and RNA. In this manner, anaerobic glycolysis provides cells with all the requirement for mitosis, both physical (sufficiant cell mass) and biochemical (suitable molecular species).

Cancer and its hallmark, unrestrained cell growth, cannot be defined solely to a disease of the genome. Why do cell divide? To understand human cell division, a detour in the world of microbiology is necessary.

Bacteria and yeast multiply if food is abundant and if the milieu is favorable. In the living world, energy passes from the sun to photosynthetic organisms and then to other organisms in the form of the chemical bonds of sugars, lipids, and proteins. To obtain energy in a usable form, yeast must have an electron (or hydrogen) donor, which serves as an initial energy source within the cell. The nutrition, such as glucose, provides the initial electron. In aerobic respiration, oxygen serves as the final electron acceptor. In anaerobic respiration (fermentation), inorganic substances other than oxygen, such as nitrate ions (NO_3^-), serve as the final electron acceptor.

In the presence of sufficient concentration of oxygen, glucose is completely degraded into water and carbonic gas. At lower concentrations of oxygen, there is incomplete degradation of glucose and production of waste.

The great chemist and microbiologist Louis Pasteur played a central role in proving that this conversion to ethanol required living organisms rather than a chemical catalyst. In 1861, Pasteur showed that by bubbling oxygen into yeast broth the cells could be made to stop growing but ferment vigorously – an observation later called the Pasteur effect. Yeast consumes much more glucose in the absence of oxygen than in its presence. In modern terms the Pasteur effect amounts to an activation of anaerobic glycolysis in the absence of oxygen in order to meet cellular ATP utilization requirements with the much lower efficiency of ATP production by fermentation compared to respiration. Hypoxia and its consequent anaerobic glycolysis result in yeast growth and multiplication.

During anaerobic glycolysis, there is production of waste. Part of this waste is released in the extracellular component. Some wastes produced by anaerobic glycolysis are highly valuable. As a Frenchman, I am fond of wine, a waste product of prokaryotic fermentation.

Other "waste products" of anaerobic glycolysis stay inside the cell, this is the case of amino acids, lipids, glucids, and nucleic acid (1–5). These amino acids in turn form proteins, lipids are transformed into hormones, and nucleic acid is transformed into DNA and RNA. Thus we can see that thanks to anaerobic glycolysis a cell has at its disposal everything needed for mitosis. The cell mass increases, cell division becomes possible.

Like bacteria and yeast, human cells multiply, in vitro, in Petri dishes. Just like bacterial cultures, the milieu used for human cell culture is a mixture of water, electrolytes, vitamins and, above all, fetal serum (blood is an excellent milieu for bacterial culture, as septicemia clearly shows).

Many higher animals share this property of oxygen balance with yeast. When given nutrients and oxygen, they will burn fuel quickly like a stoked fire, but when deprived of oxygen, they will reproduce by cell multiplication and division (rather than metabolize). This kind of behavior – burn fuel or divide – is common to many organisms including human cells. Differentiated human cells, like the those of the muscles, kidney, or heart, use respiration. Because there is complete degradation of glucose, there is no increased cellular mass. Lesser-differentiated cells like bone marrow stem cells use both aerobic and anaerobic glycolysis.

Modern biology has been marred by the discoveries of hundreds of growth factors. Growth factors are small polypeptides. They are waste products of partial combustion of glucose by one cell. This valuable waste is metabolized and digested by another cell. These "-growth factors" are ultimately turned into energy. Adding adequate food like glucose, amino acid, or "growth factors" to a cell culture triggers cell division.

The concept of growth factors, which was for some time limited to polypeptides, now extends to sugars and fat, such as triglycerides. All these molecules have one thing in common: they deliver energy to the cell. There are no "signaling molecules" which cannot be transformed into energy. Like growth factors, hormones can be secreted by distant cells or utilized on the spot. Steroid hormones are derivatives of cholesterol, that is, of fat.

Chemists understand the importance of cellular energy. They modify "growth factors" and synthesize false nutrients, which cannot be digested and transformed into available energy for the cell. As a result, cell metabolism is reduced. If the cells select this type of "growth factor" for nourishment, they will ultimately starve to death.

Like a motor, the cell consumes energy. Its nutrients are called lipids, proteins, and sugars by some (dieticians, endocrinologists, bacteriologists), cytokines, chemokines, and growth factors by others. Cellular ecologists call these growth factors a resource.

But mitosis is not always possible. Energy and increased cell size are not sufficient to displace the rigid borders of the Petri dish. Even when it has filled up on growth factors, the cell cannot divide. Only the death of its neighbor, which diminishes constraint, allows further cell division to compensate for the loss.

Orthopedics confirms the role of constraints on cell division.To lengthen a bone, orthopedists make an incision in the bone, then exert traction. Constraint is diminished at the level of the surgical bed, and cell multiplication becomes possible. Similarly, human tendon is elongated by the application of load at its extremity (6).

Prior to a difficult operation, because of limited skin for wound closure, a skin-stretching device (an inflatable balloon) is inserted under the skin. After incremental traction the skin is elongated, the wound can be closed. Similarly, stretching of peripheral nerves increases the length of the nerve (7). Mechanical deformation of fetal rat lung cells simulating fetal respiratory movements increases cellular replication (9). Rhythmic deformation also increases the intestinal epithelial cell proliferation in a frequency-dependent manner (6, 10).

The effect of shear stress on the endothelial cell is another well-studied example (6, 8). Shear stress is responsible for vascular network formation, the fractal organization of the arterial and venous trees, as well as the unavoidable tropism of arteries toward capillaries and then veins (8). At the cellular level, shear stress induces cell proliferation. The bone, the cartilage, the nerves, the endothelium, the intestine or the lung are not unique in responding to external forces; virtually all cells are affected by the mechanical environment (6–15).

Another example of the role of physical constraints is the change in cell growth in eukaryotic cells exposed to decreased gravity during space missions like Skylab (16). The importance of physics in regulating cell growth is also confirmed by modern molecular biology techniques. Noncommunicating cells can multiply. When transfected with expressible gap-junction genes, these cells had restored cell-to-cell communication and normal constraints resulting in growth arrest (17, 18).

References

1. Gerasimovskaya EV, Ahmad S, White CW, Jones PL, Carpenter TC, Stenmark KR (2002) Extracellular ATP is an autocrine/paracrine regulator of hypoxia-induced adventitial fibroblast growth. Signaling through extracellular signal-regulated kinase-1/2 and the Egr-1 transcription factor. J Biol Chem 277: 44638–44650
2. Wada Y, Sugiyama A, Yamamoto T, Naito M, Noguchi N, Yokoyama S, Tsujita M, Kawabe Y, Kobayashi M, Izumi A, Kohro T, Tanaka T, Taniguchi H, Koyama H, Hirano K, Yamashita S, Matsuzawa Y, Niki E, Hamakubo T, Kodama T (2002) Lipid accumulation in smooth muscle cells under LDL loading is independent of LDL receptor pathway and enhanced by hypoxic conditions. Arterioscler Thromb Vasc Biol 22:1712–1719
3. Wagenmakers AJ (1992) Amino acid metabolism, muscular fatigue and muscle wasting. Speculations on adaptations at high altitude. Int J Sports Med 1:S110–S113
4. Palmer GC (1985) Cyclic nucleotides in stroke and related cerebrovascular disorders. Life Sci 36:1995–2006
5. Dickerman RW, Li HY, Wade GN (1993) Decreased availability of metabolic fuels suppresses estrous behavior in Syrian hamsters. Am J Physiol 264: R568–R572
6. Schwartz L, Maitournam H, Stoltz C, Steayert JM, Ho Ba Tho MC, Halphen B (2003) Growth and cellular differentiation: a physico-biochemical conundrum? Med Hypothesis 61:45–51
7. Hall GD, Van Way CW, Kung FT, Compton-Allen M (1993) Peripheral nerve elongation with tissue expansion techniques. J Trauma 34:401–405
8. Fleury V, Schwartz L (1999) D.L.A. from shear stress as a simple model of vasculogenesis. Fractals 7:33–39
9. Liu M, Xu J, Transwell AK, Post M (1994) Inhibition of mechanical strain-induced fetal rat lung cell proliferation by gadolinium, a stretch-activated channel blocker. J Cell Physiol 161:501–507
10. Schwartz L, Balosso J, Baillet F, Brun B, Amann JP, Sasco A (2002) Cancer, the role of extracellular disease: a hypothesis. Med Hypothesis 58:340–346

11. Murnin M, Kumar A, Li G, Brown M, Sumpio BE, Basson MD (2000) Effects of glutamine isomers on human(Caco-2)intestinal epithelial proliferation, strain responsiveness and deformation. J Gastrointes Surg 4:435–442
12. Bard J (1990) Morphogenesis. Cambridge University Press (UK)
13. Oster GF, Murray JD, Harris AK (1985) A model for chondrogenic condensations in the developing limb: the role of extracellular matrix and cell tractions. J Embryol Exp Morph 89:93–112
14. Googdman S, Aspenberg P (1993) Effects of mechanical stimulation on the differentiation of hard tissue. Biomaterials 14:563–568
15. Carter DR, Beaupre GS, Giori NJ, Helms JA (1998) Mechanobiology of skeletal regeneration. Clin Orthop 355S:41–55
16. Hughes-Fulford M (1991) Altered cell function in microgravity. Exp Gerontol 26:247–256
17. Zhu D, Caveney S, Kidder GM, Naus CC (1991) Transfection of C6 glioma cells with connexin 43 cDNA: analysis of expression, intercellular coupling, and cell proliferation. Proc Natl Acad Sci USA 88:1883–1887
18. Mehta PP, Hotz-Wagenblatt A, Rose B, Shalloway D, Loewenstein WR (1991) Incorporation of the gene for a cell-cell channel protein into transformed cells leads to normalization of growth. J Membr Biol 124:207–225

Extracellular Constraints Regulate Cell Differentiation

CHAPTER 4

The aim of this chapter is to suggest that cell differentiation, like cell multiplication, may be simply the result of physical and chemical constraints.
I take the informative example of ossification: The study of bone age, obtained from X-rays of the hand, shows that ossification of the cartilage is very precise both in time and in space. Ossification (that is differentiation of cartilage into bone) results from a combination of mechanical and biochemical constraints. This example of cell differentiation is not the only one to be explained by changes in local constraints.

Currently, the predominant hypothesis explains cellular differentiation as an essentially genetic intracellular process. But, after the deciphering of the human genome, molecular biology has yet to provide a clear picture of how and why a cell may change its destiny.

The growth and maturation of the hand is a highly predictable phenomenon (1). During embryogenesis, bones of the diaphysis are formed on an initial cartilaginous model. Cartilage is later replaced by bone; this process is called enchondral ossification (2). The long bones of the hand end near the joint in a separate epiphysis, which is ossified during childhood. The epiphysis becomes fused with the shaft of the bone (diaphysis) at puberty when growth comes to an end (2). The epiphyses undergo a characteristic series of events: central calcification, absorption of cartilage and enchondral ossification (2). At birth, there is no limited detectable bone on X-rays of the epiphysis of the hand. Bone development assessed by a simple radiological picture of the right hand has been used to confirm the age of the infant (1), assess its nutritional status if there is a delay in ossification because of an underlying disease (3), or even estimate the age of the cadaver of an immature child (4).

Coordinated Development of the Hand

The proper development of the musculoskeletal system requires the coordinated development of cartilage, bone, muscle, tendon, and cartilage (5). In the embryo, ossification of the cartilaginous anlagen of the metatarsus starts in parallel with active movement of the feet by muscle contraction (6). Mechanical stress resulting from muscle contraction seems to guide enchondral ossification patterns (6–8).

Comparison of muscle and tendon reveals that much of their morphogenesis is temporally and spatially closely associated (5). Reciprocal muscle–tendon interactions are necessary for correct muscle–tendon patterning and morphogenesis. Tendon development was examined in muscleless limbs produced by coelomic grafting of early limb buds and muscle development was analyzed in limbs where tendon had been surgically altered (5). Subdivision of muscle masses and segregation of tendon primordia into individual tendons require reciprocal interaction between muscle and tendon (5).

Cartilage Defects Are Responsible for Multiple Malformations

Achondroplasia is a rare genetic disease. The cartilage shows morphological and biochemical abnormalities frequently caused by the substitution of arginine for glycine in the fibroblast growth factor receptor (9–12). At birth, short stature and squared pelvis are not obvious (12). The deformity increases with age (12) and is not limited to the cartilage but also to the bone, the muscles, and the peripheral nerves. In more general terms, when there is bone malformation, muscles and tendons do not develop normally; their development is linked to that of the bones, probably due to changes of constraint on muscles and tendons.

Growth Plate Arrest Stops the Growth of the Hand

The use of rigid fixation for fracture of the extremity is commonplace (14). Epiphyses plated for 1 year show increased bone differentiation, premature closure, and growth arrest (14). They also have shortened muscles, tendons, and nerves. In young cancer patients, growth plate injury often results from surgery or radiation therapy. This translates into marked deformity, sclerotic metaphyseal bands, muscular atrophy, and growth arrest (15). Here again, the muscles, tendons, and nerves are shortened. Similar treatments that were delivered sparing the growth plate had markedly fewer side effects (15).

It is at puberty that growth plates are fused and growth stops. This may be due to a direct hormonal effect. But the normal growth plates are under increased pressure; the largest load comes from muscle (16). The muscle load increases around puberty. The most intense force development occurs between 12 and 15 years of age in boys, and earlier in girls (17). Girls stop growing earlier than boys do. Treatment with testosterone of young boys prior to puberty increases muscle mass and causes growth cessation (18). Similarly, heavily trained gymnasts (19) or swimmers (20) experience attenuated growth during their years of training followed by catch-up growth during a reduced training schedule or in the months following retirement.

Growth Plate Growth May Be Enough to Explain the Expansion of the Hand

Growth of long bones occurs at the growth plate, a cartilage structure that contains three main layers: the resting, proliferative, and hypertrophied zones. Growth of long bones occurs at the growth plate, a layer of cartilage that separates the epiphysis from the metaphysis. Growth plates exhibit spatial polarity. Proliferative chondrocytes undergo terminal differentiation when they approach the metaphyseal, but not the epiphyseal, border of the growth plate (21). By contrast, the elongation of the muscles, the tendons, the skin, or the nerves appears secondary to bone growth.

Muscle is highly responsive to changes in functional demands. Overload leads to hypertrophy, whereas decreased load force generation and immobilization, with the muscle in the shortened position, leads to atrophy (22).

Human tendon is elongated by the application of load at its extremity (23). Mechanical tension increases the number of fibroblasts (24).

Ossification Appears to Result from Growth Plate Growth

In 1911, Gebhardt (25, 26) suggested that the ossification of the chondroepiphysis started at a point where the accumulation of stress was the greatest. The ossification center of the epiphysis is formed by hypertrophied chondrocytes. In a short period of time, calcium phosphate is deposited in the matrix around these cells (26). In vitro, mouse metatarsals are exposed to external load (26). Intermittent ambient hydrostatic pressure increases the calcification (25).There is accelerated osteogenesis in the area of intermittent high shear (25, 27).

Recently, Maitournam (28) reviewed a data bank of X-rays of the hand of children of various ages. Using finite-element analysis he showed that the localization and the pattern of growth of ossification takes place in areas of high hydrostatic pressure and shear stress.

The fact that the combination of shear stress and pressure is correlated to mineralization appears to be a general phenomenon. The ossification of the vertebral body starts at its center, where shear stress is the greatest. Early ossification of the diaphysis of the femur takes place at a time when the joints are flatter (29, 30), again in an area of high shear. Cartilage that does not ossify (e.g., joint, nose, larynx, ear, bronchus) is not exposed to high shear stress.

Mechanical Stress Induces Cellular Differentiation

Mesenchymal stem cells are multipotent cells that can be induced to differentiate into a variety of mesenchymal tissues, including bone, cartilage, tendon, fat, bone marrow stroma, and muscle (4). Several mesenchymal cells (mechanocytes), e.g., osteoblasts and fibroblasts as well as muscle cells, are activated by mechanical strain (31). In the past two decades, it has been well established that many cells are sensitive to mechanical forces and can change their phenotype and surrounding extracellular matrix (ECM) in response to the mechanical environment (32). Traction appears to generate condensation and maturation of chondrocytes or feather, scale, and hair formation (33–39).

Chondrocytes are known to sense and respond to mechanical stimuli (40). Fluid-induced shear causes chondrocytes to elongate and align (41). Chondrocytes respond to shear stress by an increased secretion of extracellular matrix, namely collagen and proteoglycan (40), as well as the modification of metabolism (41). However, primary adult bone cells do not appear to respond to fluid-flow-induced shear stress in these physiological ranges (42).

External load also plays a critical role in determining muscle mass and its phenotype in myocytes (43). Myocytes have the ability to sense mechanical stretch and convert it into intracellular growth signals, which leads to hypertrophy. Stretch is, by itself, an important mechanical signal for the production of more actin and myosin filaments, and for the addition of new sarcomeres. This is preceded by upregulation of transcription of the appropriate genes, some of which, like the myosin isoforms, markedly change muscle phenotype. Indeed, the switch in the expression induced by mechanical activity of myosin heavy chain genes, which encode different molecular motors, is the means by which the tissue adapts to a given type of physical activity. Mechanical stretch of myocytes in vitro causes activation of multiple second messenger systems that are very similar if not identical to growth factor-induced cell signaling systems.

Similarly, traction induces the secretion of extracellular matrix by fibroblasts, distorts collagen gels and creates patterns similar to tendon (44). This morphogenetic rearrangement of extracellular matrix is the primary function of fibroblast traction and explains its excessive strength (44).

Pathologic Calcification and Ossification Can Also Be Induced by Hypoxia and Alkalosis

Circumstantial evidence suggests that hypoxia and\or alkalosis also play a key role in pathologic calcification and ossification.

Cardiac ligature, subcutaneous implantation of glass diaphragms or exposure of a transplanted tendon to anoxia results in transient chondrogenesis followed by enchondral ossification (45).

Similarly, Busher (46) reports the case of a patient who underwent gastric tube reconstruction. Following cardiac arrhythmia, he developed hypoxia and ischemic necrosis over 5 cm of the proximal gastric tube. Three weeks later that area was ossified; trabecular bone was present along the entire length of the constricted gastric tube.

Ectopic ossification has been associated with several conditions in both neoplastic and non-neoplastic tissue (46–49). Paraplegic patients frequently develop ectopic ossification (46, 50). But there is no significant difference of ectopic bone formation between paraplegic rabbits and nonparaplegic rabbits under the same immobilization and passive movement of the posterior legs. For Izumi (50), the reason may simply be poor oxygenation because of blood stasis.

Cartilage is practically avascular (nutrients are transported from the synovial fluid and vascularized subchondral bone) (62). Oxygen tensions within cartilage are therefore significantly lower than vascularized tissues ranging from 2.7% to 7.5% oxygen (62–64). When pregnant mice are exposed to hypobaric oxygen in vivo, the embryos suffer from major scoliosis. The malformations are present in the cartilaginous stage of development of the vertebral column (65).

The effects of oxygen on cell differentiation are confirmed in vitro. Oxidative stress regulates cell function and proliferation (66). Low oxygen tension promotes and induces redifferentiation of dedifferentiated bovine articular chondrocytes (67).

Calcium mineral deposition in the atherosclerotic plaque also results from hypoxia. The vascular cell, when exposed to hydrogen peroxide, differentiates into an osteoblastic cell (51); this effect is counteracted by antioxidants. Oxidative stress modulates the differentiation of bone and vascular cells in opposite ways (51).

Metabolic acidosis increases calcium efflux from bone and hypercalciuria (52). Metabolic acidosis increases osteoporosis and osteomalacia and the propensity to develop kidney stones (53). By contrast, alkalosis neutralizes endogenous acid production and improves bone mineral accretion (52, 54).

Calciphylaxis is a rapidly developing fatal process of vascular calcium deposition with prominent cutaneous manifestation. Metastatic pulmonary deposition is a complication of renal failure. The pathologic deposition of calcium is favored by alkalosis (55, 56).

Shear Stress and Pressure Modify the Extracellular Matrix and Change the Availability of Nutrients and Oxygen

How shear stress and pressure can induce cell differentiation is a matter of speculation. It may be a direct effect, the release of a differentiation factor (57) or decreased availability of oxygen because of limited diffusion due to increased extracellular matrix.

Collagen is a vital component of the extracellular matrix both of muscles and tendons (58). It acts as a scaffold to maintain muscle shape and permit even distribution of force, and plays a crucial role in the mechanical properties of the tendons. Under normal circumstances, collagen is continually being synthesized and degraded throughout life. Increased mechanical stress, which causes muscle hypertrophy, stimulates collagen synthesis. The concentration of other extracellular molecules (e.g., laminin, fibronectin, vitronectin) is also modulated by mechanical stress (59).

Mechanical compression and tension generated by the condensing mesenchyme in the limb bud (60) appear to be enough to constrict and close off the thin-walled undifferentiated blood vessels caught in the condensation foci (60), thus leading to avascular areas. The differentiation of muscle and cartilage has been interpreted in terms of vascular prepattern. Extracellular matrix and\or low oxygen tension differentiates stem cells into chondrocytes (61).

Hypothesis: Shear Stress Induces Alkalosis Which Is Known to Increase Calcium Deposition

Stylophora pistillata is a scleractinian coral. Its calcification is a function of mechanisms that concentrate the CO_2 in coral cells (68). Pancreatitis stones, a frequent complication of chronic pancreatitis, are made of calcite ($CaCO_3$). The precipitation of $CaCO_3$ is a function of the pH and the availability of bicarbonates (69). Similarly, extracellular alkalosis increases calcium precipitation at the periphery of pancreatic B cells (70).

The pH surrounding the osteoclast is highly acidic (71). The apical bone-resorbing compartment of the osteoclast is sealed off by the attachment of the osteoclast to the calcified matrix and is actively acidified by the osteoclast. In the low pH environment of the bone-resorbing lacuna produced by the osteoclast, the mineral phase dissolves, exposing the organic matrix to the action of the secreted enzymes. These observations are consistent with a scheme in which, in the low pH environment of the bone-resorbing lacuna produced by the osteoclast, the mineral phase dissolves, exposing the organic matrix to the action of the secreted enzymes. The activity of these enzymes is in turn presumably favored by the acidic milieu. All constituents of the matrix, whether mineral or organic, would then be reduced to their elemental forms (ions and amino acids) extracellularly.

The secretion of acid by proton pumps involves cation-independent mannose-6-phosphate receptors. These receptors bind to an enzyme-linked mannose-6-phosphate (71). There is "in vitro" evidence of the importance of pH in bone formation and resorption. Metabolic acidosis increases urinary calcium excretion (54). By contrast, alkalosis neutralizes endogenous acid production and improves bone mineral accretion (54). Alkalosis causes a decrease in osteoclastic beta-glucuronidase release and an increase in osteoblastic collagen synthesis and calcium deposition on the bone cells (72).

The link between hypoxia, alkalosis, and calcium precipitation is complex and beyond the realm of this book. Chemical hypoxia induces hypocapnia alkalosis in primary culture (73). Similarly, ischemia and hypoxia may induce alterations of ion homeostasis, including alkalosis (74).

Conclusion

Currently, the predominant hypothesis explains cellular differentiation as an essentially genetic intracellular process. However, it is possible or even probable that this differentiation is the result of simple biochemical and physical events of extracellular origin. These changes trigger complex modifications in gene expression, which in turn modulate cell differentiation.

References

1. Graham CB (1972) Assessment of bone maturation-methods and pitfalls. Radiol Clin North Am 10:185–202
2. Cancedda R, Castagnola P, Cancedda FD, Dozin B, Quarto R (2000) Development control of chondrogenesis and osteogenesis. Int J Dev Biol 44:707–714
3. Sutphen JL (1985) Growth as a measure of the nutritional status. J Pediatr Gastroenterol Nutr 4:169–181
4. Ubelaker DH (1987) Estimating age of death from immature human skeletons: an overview. J Forenscic Sci 32:1254–1263
5. Kardon G (1998) Muscle and tendon morphogenesis in the avian hind limb. Development 125:4019–4032
6. Burger EH, Klein-Nulend J, Veldhuizen JP (1991) Modulation of osteogenesis in fetal rudiments by mechanical stress. J of Biomechanics 24:101–109
7. Carter DR, Wong M, Orr TE (1991) Musculoskeletal ontogeny, phylogeny, and functional adaptation. J Biomechanics 24 S1:3–16
8. Turner C, Pavalko F (1998) Mechanotransduction and functional response of the skeleton to physical stress: the mechanism and mechanics of bone adaptation. J Orthop Sci 3:346–355
9. Miura T (1981) Congenital hand anomalies, and their association with other congenital abnormalities. Hand 13:267–270
10. Stanescu V, Stanescu R, Maroteaux P (1984) Pathogenic mechanisms in osteochondrodysplasia. J Bone Joint Surg Am 66:817–836
11. Hall BK (1981) Intracellular and extracellular control of the differentiation of cartilage and bone. Histochem 13:599–614
12. Matsui Y, Kawabata H, Ozono K, Yasui N (2001) Skeletal develpment of achondroplasia. Pediatr Int 43:361–363
13. Roberts A,Evans GA (1993) Orthopedic aspects of neuromuscular disease in children. Curr Opin Pediatr 5:379–383

14. Opperman LA, Drake DB, Kang P, Nakeeb SM, Gammas M, Morgan RF (2000) A pilot study of short and long term sequelae to rigid fixation across metaphyseal physes in a baboon model. Ann Plast Surg 44:36–43
15. Roebuck DJ (1999) Skeletal complications in paediatric oncology. Radiographics 19:873–885
16. Schiessl H, Frost HM, Jee WS (1998) Estrogen and bone-muscle strength and mass relationships. Bone 22:1–6
17. Backman E (1988) Methods for measurement of muscle function. Methodological aspects, reference values for children, and clinical applications. Scand J Rehabil Med Suppl 20:9–95
18. Soliman AT, Khadir MM, Asfour M (1995) Testosterone treatment in adolescent boys with constitutional delay of growth. Metabolism 44:1013–1015
19. Caine D, Lewis R, O'Connor P, Howe W, Bass S (2001) Does gymnastics training inhibit growth of females? Clin J Sport Med 11:260–270
20. Wawryniak G (2001) Biological age in children who practise swimming. Anthropol Anz 59:149–156
21. Abad V, Uyeda JA, Temple HT, De Luca F, Baron J (1999) Determinants of spatial polarity in the growth plate. Endocrinology 140:958–962
22. Goldspink G (1999) Changes in muscle mass and phenotype and the expression of autocrine and systemic growth factors by muscle in response to stretch and overload. J Anat 194:323–334
23. Maganaris CN (2000) Load-elongation characteristics of in vivo human tendon and aponeurosis. J Exp Biol 203:751–756
24. Breen EC, Fu Z,Normand H (1999) Calcyclin gene expression is increased by mechanical strain in fibroblasts. Am J Respir Cell Mol Biol 21:746–752
25. Burger E, Klein-Nulend J, Veldhuuzeen JP (1991) Modulation of osteogenesis in fetal bone rudiments by mechanical stress in vitro. J Biomechanics 24:101–109
26. Tanck E, Van Driel WD, Hagen JW, Burger EH, Blankevoort L, Huiskes R (1999) Why does intermittent hydrostatic pressure enhance the mineralization process in fetal cartilage? J of Biomechanics 32:153
27. Strayer L (1971) Embryology of the human hip joint. Clinical Orthopaedics and Related Res 74:221–240
28. Maitournam H, Chassignet P, Halphen B, Schwartz L. Shear stress and ossification. CR Acad Sci (Paris) submitted
29. Bard J (1990) Morphogenesis. Cambridge University Press (UK)
30. Shi-Jun L, Wei-Dong L, Rui-Xue M, En-Guang K (1998) Morphometric study on development of fetal hip. Chinese Medical J 101:750–754
31. Caterson EJ, Nesti LJ, Albert T, Danielson K, Tuan R (2001) Application of mesenchymal stem cells in the regeneration of musculoskeletal tissue. Med Gen Med E1

32. Swartz MA, Tschumperlin DJ, Kamm RD, Drazen JM (2001) Mechanical stress is communicated between different cell types to elicit matrix remodelling. P.N.A.S. 98:6180–6185
33. Fleury V, Schwartz L (1999) D.L.A. from shear stress as a simple model of vasculogenesis. Fractals 7:33–39
34. Liu M, Xu J, Transwell AK, Post M (1994) Inhibition of mechanical strain-induced fetal rat lung cell proliferation by gadolinium, a stretch-activated channel blocker. J Cell Physiol 161:501–507
35. Murnin M, Kumar A, Li G, Brown M, Sumpio BE, Basson MD (2000) Effects of glutamine isomers on human(Caco-2)intestinal epithelial proliferation, strain responsiveness and deformation. J Gastrointes Surg 4:435–442
36. Bard J (1990) Morphogenesis. Cambridge University Press (UK)
37. Oster GF, Murray JD, Harris AK (1985) A model for chondrogenic condensations in the developing limb: the role of extracellular matrix and cell tractions. J Embryol Exp Morph 89:93–112
38. Googdman S, Aspenberg P (1993) Effects of mechanical stimulation on the differentiation of hard tissue. Biomaterials 14:563–568
39. Carter DR, Beaupre GS, Giori NJ, Helms JA (1998) Mechanobiology of skeletal regeneration. Clin Orthop 355S:41–55
40. Jim M, Franck EH, Quinn TM, Hunziker EB, Grodzinsky AJ (2001) Tissue shear deformation stimulates proteoglycan and protein biosynthesis in bovine cartilage explants. Arch Biochen Biophys 395:41–48
41. Smith RL, Donlon BS, Gupta MK, Mohtai M, Das P, Carter DR, Cooke J, Gibbons G, Hutchinson N, Schurman DJ (1995) Effects of fluid induced shear stress on articular chondrocytes morphology and metabolism in vitro. J Orthop Res 13:824–831
42. Brand RA, Stanford CM, Nicolella DP (2001) Primary adult bone cells do not respond to tissue (continuum) level strains. J Orthop Sci 6:295–301
43. Sadoshima J, Izumo S (1997) The cellular and molecular response of cardiac myocytes to mechanical stress. Annu Rev Physiol 59:551–571
44. Harris AK, Stopak D, Wild P (1981) Fibroblast traction as a mechanism for collagen morphogenesis. Nature 290:249–251
45. Danis A (1992) Soft tissue ossification: mechanism. Bull Mem Acad R Belg 147: 298–306
46. Buscher HC, Van Lanschot JJ, Mulder AH, Tilanus HW (1995) Heterotopic ossification induced by hypoxia in a retrosternal gastric tube following transhiatal. Oesophagectomy. J Clin Pathol 48:177–178
47. Ackerman NB (1968) Bone formation in the gastric mucosa following subtotal gastrectomy in rats. Exp Mol Pathol 9:125–130
48. An HS, EbrahemN, Kim K,Jackson WT, Kane JT (1987) Heterotopic ossification and pseudoarthrosis in the shoulder following encephalitis. Clin Orthop 219: 291–298

49. Asa DK, Bertorini TE, Pinals RS (1986) Myositis ossificans circumscripta: a complication of tetanus. Am J Med Sci 292:40–43
50. Izumi K (1983) Study of ectopic bone formation in experimental spinal cord injured rabbits. Paraplegia 21:351–363
51. Mody N, Parhami F, Sarafian TA, Demer LL (2001) Oxydative stress modulates differentiation of vascular and bone cells. Free Rad Biol Med 31:509–519
52. Bushinsky DA, Krieger NS, Geisser DI, Grossman EB, Coe FL (1983) Effects of ph on bone calcium and protons flux in vitro. Am J Physiol 245:204–209
53. Wiederkehr M, Krapf R (2001) Metabolic and endocrine effects of metabolic acidosis in human. Swiss Med Wkly 131:127–132
54. Bushinsky DA (1996) Metabolic alkalosis decreases bone calcium efflux by suppressing osteoclasts and stimulating osteoblasts. Am J Physiol 271:216–222
55. Matsuo T, Tsukamoto Y, Tamura M, Hanakoa M, Nagaoka T, Kobayashi Y, Higashiara M, Yokoyama H, Saegusa N (2001) Acute respiratory failure to "pulmonary calciphylaxis" in a maintenance haemodialysis patient. Nephron 87: 75–79
56. Bloodworth J, Tomashefski JF (1992) Localised pulmonary metastatic calcification associated with pulmonary artery obstruction. Thorax 47:174–178
57. Jin M, Frank EH, Quinn TM, Hunziker EB, Grodzinsky AJ (2001) Tissue shear deformation stimulates proteoglycan and protein biosynthesis in bovine cartilage explants. Arch Bioch Biop 395:41–48
58. Bishop JE (1998) Regulation of cardiovascular collagen deposition by mechanical forces. Mol Med Today 4:69–75
59. Loeser RF (2000) Chondrocyte integrin expression and function. Biorheology 37:109–116
60. Wilson DJ (1986) Development of avascularity during cartilage differentiation in the embryonic limb. Differentiation 30:183–187
61. Von Der Mark K, Conrad G (1979) Cartilage and cell differentiation: a review. Clin Orthop 139:185–205
62. Murphy C, Sambanis A (2001) Effect of oxygen tensionon chondrocyte extracellular matrix accumulation. Connective Tissue Res 42:87–96
63. Brighton CT, Heppenstall RB (1971) Oxygen tension in zone of the epiphyseal plate the metaphysis and the diaphysis. Am J Bone Joint Surg 53:719–728
64. Ellis S, Velayutham M, Velan S, Petersen E, Zweiler J, Kuppusamy P, Spencer R (2001) EPR oxygen mapping of engineered cartilage in reactor grown in a hollow-fiber bioreactor. Mag Res Med 46:819–826
65. Rivard CH, Labelle P, Simoneau R, Duhaime M, Poitras B, Simard P (1982) Moderate hypobaric hypoxia used as an inducer of congenital vertebral malformation in mouse's embryo. Chir Pediatr 23:65–67
66. Butler AJ, Eagleton MJ, Wang D, Howell RL, Strauch AR, Khasg iwala V, Smith HC (1991) Induction of the proliferative phenotype in differentiated myogenic cells by hypoxia. J Biol Chem 266:18 250–18 258

67. Domm C, Schunke M, Christesen K, Kurz B (2002) Redifferentiation of dedifferentiated bovine articular chondrocytes in alginate culture under low oxygen tension. Osteoarthritis Cartilage 10:13–22
68. Furla P, Galgani I, Durand I, Allemand D (2000) Sources and mechanisms of inorganic carbon transport for coral calcification and photosynthesis. J Exp Biol 203:3445–3457
69. Moore EW, Verine HJ (1981) Pancreatic calcification:formation constants of CaHCO3+ and Ca CO3(0) complexes determined with CA2+ electrodes. Am J Physiol 241:G182–190
70. Rebolledo OR, Semino MC, Gaglardino JJ (1988) Effect of extracellular alkalosis upon calcium distribution within B cells. Acta Physiol Latinoam 38: 329–343
71. Baron R (1989) Molecular mechanisms of bone resorption by the osteoclast. Anat Rec 224:317–324
72. Carano A, Schlesinger PH, Athanasou NA, Teitelbaum SL, Blair HC (1993) Acid and base effects on avian osteoclast activity. Am J Physiol 264:694–701
73. Jorgensen NK, Petersen SF, Damgaard I, Shousboe A, Hoffmann EK (1999) Increase in CA2+ and changes in intracellular ph during chemcal anoxia in mouse neocortical neurons in primary culture. J Neurosci Res 56:358–370
74. Katsura K, Ekholm A, Siesjo BK (1992) Coupling among changes in energy metabolism, acid-base homeostasis and ion fluxes in ischaemia. Can J Physiol Pharmacol 70:S170–S175

Changes in Cellular Constraints During Carcinogenesis

CHAPTER 5

Cancer is considered a multistep process in which multiple genetic alterations occur that have a cumulative effect on the control of cell proliferation, division, and growth. This view of cancer, however, is not exclusive. In this chapter, we review supporting evidence that carcinoma may be secondary to the disorganization of an organ and should not be considered as a disease limited to the epithelial cell. This disorganization may have multiple causes, such as embryonic remnants, severe burns, or, more often, inflammation and chronic fibrosis.
This disorganization results in change of physical and biochemical constraints.

In the previous chapters, we saw that cell growth and differentiation are controlled by external physical and biochemical constraints. In this chapter, we review supporting evidence that carcinogenesis (i.e., the induction of cancer) is a direct consequence of the changes in the architecture of the targeted organ. This tissue disruption results in decreased mechanical constraints allowing for loss of cell differentiation, mitosis, and proliferation.

Disorganization of the Organ Precedes Epithelial Tumor Formation

Liver Cancer

Cancer is frequently associated with preexisting tissue disorganization or disruption. Between 60% and 90% of hepatocellular carcinoma occurs in patients with hepatic macronodular cirrhosis (1, 2). Chronic liver disease of any type is a risk factor for liver cancer. The cancer may be caused by hepatitis C or B, alcoholic liver disease, antitrypsin deficiency, hemochromatosis, and tyrosinemia. Its features result from hepatocyte necrosis, extensive fibrosis, connective tissue deposition, vascular distortion and nodular regeneration of the remaining tissue parenchyma (2). Evidence for a cause–effect link between cirrhosis and hepatocarcinoma is lacking. The relationship may often be one of chance alone, since not all cirrhotics develop cancer. Nonetheless, diseases that cause cirrhosis also increase the risk of hepatocarcinoma (2). Furthermore, the more disorganized the liver becomes, the higher the risk of hepatocarcinoma (2, 3).

Lung Cancer

Similarly, lung cancer is most common among patients suffering from any form of chronic lung disease (4, 5). History of chronic bronchitis, emphysema, primary lung fibrosis, chronic lung infection, and even lung irradiation is associated with increased cancer risk (4, 5). There is no evidence for a relationship between bronchitis or emphysema and lung cancer that could not be explained by independent links to exposure to tobacco smoke or other noxious agents. Nevertheless, the risk of lung cancer increases with the extent of disruption of normal lung architecture. For example, the risk of lung cancer is higher among patients suffering from chronic bronchitis and severe impairment of the carbon monoxide diffusing capacity of the lungs than in those with normal lung function (6).

Breast Cancer

Breast cancer genesis also seems to be linked to architectural changes. A woman's reproductive history is one of the most important determinants of breast cancer risk. This is not a new notion. Ramaziani in 1700 first showed that breast cancer risk was higher among nuns (7). Early in the 1900s, investigations noted that nulliparity and a history of never having breast-fed an infant were risk factors. Modern epidemiological cohorts have confirmed the increased risk for breast cancer after early puberty, late menarche, and hormonal stimulation (7). In 1977, Ing reported a disproportionate increase of cases of postmenopausal breast cancer in the left breast of Tanka women of Hong Kong thought to have nursed only with the right breast (8). All these risk factors and others, like radiation to the developing breast, are related (causally or not) to change in the architecture of the breast.

Breast cancer is rare among young women. It is before the menopause, when the architecture of the mammary gland starts to undergo fatty tissue involution, that the incidence of cancer rises (7). With the completion of menopause, the breast changes, it becomes somewhat smaller and less dense. There is a decrease in the number and size of the ducts. These atrophic lobules are seen lying in a dense fibrous matrix. Increase in the connective tissue is a prominent feature of this aging process (7).

The rare cases of breast cancer in young women are often due to hereditary anomalies. The most studied gene is BRCA-1. BRCA-1 is a nuclear phosphoprotein expressed in a broad spectrum of tissues during cell division. The inheritance of a mutant BRCA-1 allele dramatically increases a woman's lifetime risk for developing both breast and ovarian cancers. This increased risk may be secondary to architectural changes. Analysis of cases of prophylactic subcutaneous mastectomy after genetic counseling for either carrying the BCRA-1 gene or belonging to a pedigree with familial breast cancer shows a different architectural pattern. BCRA-1 or related genes may have a functional role in the branching pattern of the breast during lobular development, mainly in epithelial–stroma interactions (9). BCRA-1-deficient mice display multiple malformations (10).

Childhood Tumors

Cancers occurring during childhood (nephroblastoma, medulloblastoma, retinoblastoma, or Li-Fraumeni syndrome) are also associated with tissue disorganization from embryonic remnants (1, 11, 12). Patients suffering from genetically encoded hereditary tumors such as Li-Fraumeni syndrome and retinoblastoma have both mutated epithelial cells and fibroblasts with impaired growth, resulting in concomitant malformations (12, 13).

Experimental Evidence that Tissue Disruption Contributes to Cancer

Chemical Carcinogenesis

Exposure to chemical carcinogens is considered to cause most human cancers (1). In animal carcinogenesis experiments, a first chemical (initiator) is responsible for an intense inflammatory reaction. A second chemical (promotor) is genotoxic. The word "genotoxic" has been created to replace the previous term, "toxic." The effect of genotoxic compounds is not confined to the epithelial cell; they kill epithelial and stromal cells. Genotoxicity causes tissue disruption through cell killing and replacement. For example, hepatocyte necrosis induced by genotoxic compounds, which precedes hepatic carcinoma, is associated with substantial damage to surviving hepatocytes as well as extensive mesenchymal changes and loss of normal liver architecture (14).

In vitro, however, carcinogens have not always successfully transformed normal human cells in culture (15). For these normal cells to be transformed, they often need to be immortalized by transfection with a cancer-associated virus prior to exposure to a carcinogen (16).

Radiation-Induced Cancer

Ionizing radiation induces cancer in humans and animals (17, 18). In vitro, the vast majority of attempts to achieve transformation of normal human cells into cancer cells have been unsuccessful (18). In fact radiation-induced carcinogenesis appears to be a consequence of inflammation and fibrosis.

The female mammary gland is unique among all glands in that the epithelium develops after the birth from a rudiment that can be easily removed in the young rodent at about 3 weeks of age. Barcellos-Hoff irradiated the whole mammary gland of nurturing mice. After the irradiation, the epithelial cells are surgically removed and replaced with transplanted normal mammary cells. The cancer arises from these normal non-irradiated epithelial cells. Tumor growth appears as a consequence of the changes in the irradiated stroma (18).

Physical Carcinogenesis

Chemicals and radiation induce inflammation and tissue disruption. The question is whether these architectural changes cause cancer. The answer lies in the old literature on physical carcinogenesis.

It has been documented that some foreign bodies induce cancer (19–23). The carcinogenicity of foreign bodies is linked to their shape. Cellulose membrane filters of specific shape, texture, or size generate sarcoma. Intense inflammation and proliferative fibrosis precede tumor formation. The combination of shape and size (about the width of a human cell) may also be critical (23). The carcinogenicity of this chemically inert molecule is also linked to particle shape and size.

The International Agency for Research on Cancer evaluated the carcinogenic effect of surgical implants and other foreign bodies in humans (22). The evaluation resulted in a group 2B classification (possibly carcinogenic for humans) for polymeric implants prepared as thin smooth films, implanted foreign bodies consisting of metallic cobalt or nickel, and a particular alloy powder consisting of 66–67% nickel, 13–16% chromium, and 7% iron. The evaluation also resulted in a group 3 classification (not classifiable as to their carcinogenicity to humans) for organic polymeric materials as a group, orthopedic implants of complex composition, cardiac pacemakers, silicone breast implants, dental materials, and ceramic implants.

Physical carcinogenesis may also be a "transforming" factor. In nude mice, the implantation both of colon adenoma cells and of a plastic plate is necessary for tumorigenic growth. Again, locally, there is intense inflammation (24).

Conclusion

In humans, inflammation and its consequent tissue disruption are major risk factors for subsequent cancer. This inflammation appears to have multiple causes: infection, chemical, physical, aging. In the subsequent chapters, we will see that inflammation, whatever its apparent causes, is in fact a direct consequence of anaerobic glycolysis.

References

1. De Vita V, Hellman S, Rosenberg S (1993) Cancer: principles and practice of oncology. Lippincott, Philadelphia, USA
2. Podolsky D, Isselbacher K (1994) Alcohol-related disease and cirrhosis. Principles of internal medicine. Harrisson Ed, 13th edition, 1483–1495. Mc Graw-Hill, USA
3. Baffis V, Shrier I, Sherker AH, Szilagyi A (1999) Use of interferon for prevention of hepatocellular carcinoma in cirrhotic patients with hepatitis B or C infection. Ann Intern Med 131:696–701
4. Ernster VL (1996) Female lung cancer. Annu Rev Public Health 17:97–114
5. Maitre B, Similowski T, Derenne JP (1996) Bronchopneumopathies chroniques obstructives. In: Aubier M, Fournier M, Pariente R (eds) Pneumologie. Flammarion Ed, Paris, France, pp 589–608
6. Barreto SS, Mc Lean PA, Szalai JP, Zamel N (1993) Reduction of lung diffusion for carbon monoxide in patients with lung carcinoma. Chest 103:1142–1146
7. Haagensen CD (1986) Diseases of the breast, 3rd edition. Saunders, Philadelphia, USA
8. Ing R, Hoj H, Petakis NL. Unilateral breast feeding and breast cancer. Lancet 2:124–127
9. Russo J, Lynch H, Russo IH (2001) Mammary gland architecture as a determining factor in the susceptibility of the human breast to cancer. Breast J 7:278–291
10. Cressman VL, Backlund DC, Avrutskaya AV, Leadon SA, Godfrey V, Koller BH (1999) Growth retardation, DNA repair defects,and lack of spermatogenesis in BRCA1-deficient mice. Mol Cell Biol 19:7061–7075
11. Pivnick EK, Furman WL, Velagaleti GV, Jenkins JJ, Chase NA, Ribeiro RC (1998) Simultaneous adrenocortical carcinoma and ganglioneuroblastoma in a child with Turner syndrome and germline p53 mutation. J Med Genet 35:328–332
12. Boyle JM, Spreadborough A, Greaves MJ, Birch JM, Scott D (2000) Chromosome instability in fibroblasts derived from Li-Fraumeni syndrome families. Br J Cancer 83:1136–1138
13. Le Couter JE, Kablar B, Hardy WR, Ying C, Megeney LA, May LL, Rudnicki MA (1998) Strain-dependent myeloïd hyperplasia and accelerated cell cycle in mice lacking the Rb-related p107 gene. Mol Cell Biol 18:7455–7465
14. Ames BN, Gold LS (1990) Chemical carcinogenesis. Proc Nat Acad Sci USA 87:7772–7776
15. Mc Cormick JJ, Fry DG, Hurlin PJ, Morgan TL, Wilson DM, Maher VM (1990) Malignant transformation of human fibroblasts by oncogene transfection or carcinogen treatment. Prog Clin Biol Res 340D:195–205
16. Dipaolo JA, Burkhardt A, Doniger J, Pirisi L, Popescu NC, Yamsumoto S (1986) In vitro models for studying the molecular biology of carcinogenesis. Toxicol Pathol 14:417–423

17. Rhim JS, Thraves P, Dritschilo A, Kuettel MR, Lee MS (1993) Radiation-induced neoplastic transformation of human cells. Scanning Micros 7:209–215
18. Barcellos-Hoff MH, Ravani SA (2000) Irradiated gland mammary stroma promotes tumorogenic potential by unirradiated epithelial cells. Cancer Res 60:1254–1260
19. Sonnenschein C, Soto AM (1999) The society of cells. Bios Scientific Publishers, Oxford, UK
20. Brand KG (1982) Cancer associated with asbestosis, schistomatosis, foreign body or scar cancer: A comprehensive treatise (Ed: Becker). Plenum press, New-York, 661–692
21. Stanton M, Wrench C (1972) Mechanism of mesothelioma: induction with asbestosis and fibrous glass. J Nat Cancer Inst 48:797–821
22. International Agency for Research on Cancer (1999) IARC monographs on the evaluation of carcinogenic risks to humans. Vol.74 Surgical implants and other foreign bodies. IARC Press, Lyon, France
23. Lipkin LE (1980) Cellular effects of asbestos and other fibers: correlations with in vivo induction of pleural sarcoma. Environ Health Perspect 34:91–102
24. Okada F, Kawaguchi T, Habelhah H, Koyabashi T, Tazawa H, Takeichi N, Kitagawa T, Hosokawa M (2000) Conversion of human colonic adenoma cells to adenocarcinoma through inflammation in nude mice. Lab Invest 80:1617–1628

Hypoxia Is Responsible for Changes in Cellular Constraints During Inflammation

CHAPTER 6

Since the seminal work of Warburg, it has become evident that solid human tumors are deficient in oxygen.
Inflammation appears to be a consequence of hypoxia. Hypoxia is responsible for lymphocyte infiltration, collagen deposition, and polypeptide secretion (TNF, growth factors, and interleukin) as well as blood vessel dilatation. Return to normoxia reverses these abnormalities suggesting that inflammation and transient glycolysis may be synonymous.
Glycolysis induces oncogene activation, P53 accumulation, or cytokine secretion which, in turn, mimic a situation of reduced oxygen availability, thus being responsible for further damage and tumor progression.

Inflammation was defined by the ancient Greeks as redness, pain, and heat. It can be secondary to trauma, infection, or irritation. Inflammation, fibrosis, and cancer share a common features: infiltration by leukocytes, neoangiogenesis, and intense release of cytokines and growth factors.

The goal of this chapter is to show that these features are secondary to decreased concentration of available oxygen (i.e., hypoxia).

Cancer Cells Are Hypoxic

Since the seminal work of Warburg, it has become evident that solid human tumors are deficient in oxygen (1). The presence of hypoxia has been demonstrated in cervical cancer, squamous cell carcinoma of the head and neck, melanoma, and breast cancer (2, 3). It is now widely accepted that the metabolic microenvironment of a tumor can dramatically influence a range of factors such as proliferation rate, cell cycle position, growth rate, and the development of apoptosis and necrosis (2–5).

Low oxygen tension in the primary tumor is associated with metastasis in soft tissue sarcoma, cervical carcinoma, and carcinoma of the head and neck (6). Hypoxia induces genes involved in the metastatic cascade-like matrix, such as metalloproteinase, which degrade the surrounding stroma and decrease the secretion of its inhibitors (3, 7). Similarly, cell surface integrins and other adhesion molecules are down-regulated by hypoxia (8).

For over fifty years, it has also been known that hypoxic cells are resistant to radiation therapy. When using ionizing radiation, the dose required to produce the same amount of cell killing is up to 3 times higher for hypoxic cells (5). Hypoxic cancer cells are also resistant to chemotherapy (3).

Inflammation Occurs in Hypoxic Tissues

Less research has been focused on hypoxia and inflammation. Nevertheless, there is ample evidence that inflammation occurs in hypoxic tissue (9–14). For example, episodic hypoxia during brain aging contributes to brain inflammation and Alzheimer's disease (14).

For intensive care practitioners, inflammation, and transient hypoxia are synonymous. The "cardioinflammatory response to heart failure" is an inflammation of the peripheral tissues, which results from cardiac failure (4, 15, 16). There is an infiltration by lymphocytes and monocytes and concomitant intense release of proinflammatory cytokines such as tumor necrosis factor and interleukin (4, 15–19).

Most cytokines are secreted by extramyocardial tissues and result from peripheral hypoxia. The increased level of cytokines could further reduce the endothelium-dependent vasolidator response, thereby creating a vicious cycle of more severe tissue hypoperfusion, more profound hypoxia, and more intense cytokine production (4, 19, 20).

The signs of systemic inflammatory response disappear after successful mechanical circulatory support using biventricular assist device systems. During mechanical circulatory support, elevated levels of inflammatory mediators are indicative of persistent peripheral hypoxia (4).

Hypoxia Explains Features Common to Cancer and Inflammation

Hypoxia explains features common to inflammation, fibrosis, and cancer, namely: lymphocytic infiltration, growth factor release, angiogenesis, and extracellular matrix deposition (Tables 6.1, 6.2).

Leukocyte Infiltration

Hypoxia regulates the distribution of macrophages in tumors and other inflammatory conditions. The inhibition of monocyte migration by hypoxia is rapid and reversible (21). Macrophage migration inhibitory factor (MIF) plays a pivotal role in the control of inflammatory responses. Its concentration is dramatically increased after hypoxia (22).

A hypoxic microenvironment stimulates the expression of a variety of cytokines. For example, hypoxia (2% O_2) as opposed to a normal concentration of oxygen increases the production of cytokines in human peripheral mononuclear cells (23). In hypoxia, interleukin

Table 6.1. Link between hypoxia and cancer

1) Hypoxia causes inflammation
   a. Skin redness
   b. Angiogenesis
   c. Infiltration by leukocytes
   d. Glycolysis
   e. Extracellular matrix deposition
2) Fibrosis perpetuates hypoxia
3) Hypoxia contributes to:
   a. Cell multiplication
   b. Loss of cell differentiation
   c. Increased protease secretion
   d. Metastasis

Table 6.2. Consequences of hypoxia

1) Loss of cell polarity
2) Anaerobic glycolysis
   a. DNA synthesis
   b. Secretion of growth factors and cytokines
3) Architectural changes
   a. Collagen deposits
   b. Angiogenesis
   c. Cell migration (metastasis)
   d. Stroma formation
4) Resistance to treatment
   a. Radiotherapy
   b. Chemotherapy
   c. Hormones

(IL)-2, IL-4, and interferon-gamma production is increased by 110%, 70%, and 50%, respectively.

Growth Factor Release

The primary trigger to activate expression of growth factors and their receptors appears to be hypoxia (24, 25). Hypoxia increases the expression of specific genes such as epidermal growth factor, basic fibroblast growth factor, platelet-derived growth factor, and vascular endothelial growth factor. Hypoxia also induces a marked elevation of transforming growth factor-beta gene expression (24–26).

Vasodilatation and Neoangiogenesis

It is a well-established fact that tumors but also keloid, radiation-induced fibrosis cirrhosis or chronic heart failure secrete factors that directly or indirectly stimulate angiogenesis. For example, cirrhosis consists of hepatocyte nodules surrounded by highly vascularized fibrous tissue (26). Vascular remodeling is also a constant feature of chronic heart failure.

The simple induction of hypoxia affects the expression of vascular endothelial growth factor (VEGF) and of VEGF receptors in vivo and in vitro (26).

Transgenic mice expressing constitutively active hypoxia inducible factor-1 alpha (HIF-1 alpha) in the epidermis displayed a 66% increase in dermal capillaries, a 13-fold elevation of VEGF expression, and a six- to ninefold induction of VEGF (27).

Extracellular Matrix Deposition

Growing evidence indicates that hypoxia activates extracellular matrix deposition (28, 29). Intermittent high-altitude hypoxia induces pulmonary hypertension and right ventricular hypertrophy in adult rats; prolonged hypoxia also increased the relative left ventricular mass (20).

The increased secretion of collagen during hypoxia has been reproduced in vitro (30–32). Prolyl-4-hydroxylase catalyzes the formation of 4-hydroxyproline in collagens. Analysis of the promotor region of prolyl 4-hydroxylase alpha gene shows a motif similar to the hypoxia-responsive element (HRE) of hypoxia-inducible genes such as erythropoietin (32). This O_2-dependent hydroxylase is an oxygen sensor (30). In hypoxia, the O_2 required for prolyl hydroxylation is limited (30). Hypoxia creates a molecular environment that modifies the triple helix folding of collagen (29). Thus hypoxia is responsible of extracellular collagen deposition.

Hypoxia-Inducible Factor Is Elevated During Hypoxia and Inflammation

Many molecular and physiological responses to hypoxia are controlled by the transcription factor hypoxia-inducible factor (HIF) (30). Under conditions of normoxia, HIF is virtually undetectable due to its rapid degradation. Its liability is mediated by an oxygen-dependent degradation domain whose effect is suppressed by hypoxia (33).

There is an emerging body of evidence suggesting that HIF is upregulated in inflammation, confirming the presence of hypoxia. For example, there is a marked increase in HIF during the inflammatory response of hemorrhagic shock (13). During early inflammatory events in wound healing, an increase in HIF seems responsible for the recruitment of neutrophils and macrophages to the wound site (34). Similarly, there is an overexpression of HIF by macrophages in rheumatoid synovia (35).

von-Hippel Lindau: Activation of HIF Causes Tissue Disruption

Accumulation of HIF not only occurs under conditions of hypoxia but also in von Hippel-Lindau syndrome (VHL). VHL is an autosomal, dominant inherited syndrome which predisposes carriers to changes in tissue architecture (cysts in the kidney, pancreas, epididymis, and broad ligament), angiogenesis (hemangioblastoma in the retina and central nervous system) and multiple cancers (renal cell carcinoma, pheochromocytoma) as well as polyglobulia (36–39).

In normal cells, the product of the VHL gene targets HIF for oxygen-dependent proteolysis, acting as the substrate recognition component of an E3 ubiquitin ligase (40). In VHL defective cells, this process is blocked, leading to constitutive up-regulation of HIF-1 alpha subunits, activation of the HIF complex, and overexpression of HIF target genes (40).

Hypoxia Induces Oncogene and p53 Expression, Which in Turn Mimic Hypoxia

During carcinogenesis, there is an activation of oncogenes and overexpression of the transcription factor P53. Hypoxia is responsible for an elevated mutation frequency and a mutation pattern similar to that seen in tumors (41). Hypoxia has been reported to enhance the transcription of the oncogenes c-*jun*, c-*fos*, or *neu* (40–42). Hypoxia increases p53 protein levels in normal and cancer cells (53, 54) which, in turn, leads either to cellular growth arrest at the G_1/S or G_2/M transitions of the cell cycle or to programmed cell death (apoptosis) (44, 45). Hypoxia contributes to the selection of tumor cells expressing mutations in the p53 gene partially through an increase in concentration in MIF released by T cells and macrophages (22, 48, 49). Tumor cells that develop mutations in p53 demonstrate a diminished apoptotic potential, which may contribute to further growth and tumor metastasis.

Oncogene activation, accumulation of P53 or interleukin secretion most likely stabilize HIF by inhibiting prolyl hydroxylase activity, thus mimicking a situation of reduced oxygen availability (50–53).

Conclusion

Fifty years ago, Warburg (1) stated: "Cancer cells originate from normal body cells in two phases. The first phase is the irreversible injuring of respiration. Just as there are many remote causes of plague – heat, insects, rats – but only one common cause, the plaque bacillus, there are a great many remote causes of cancer – tar, rays, arsenic, pressure, urethane – but there is only one common cause into which all other causes of cancer merge, the irreversible injuring of respiration." Modern biology appears to confirm these early findings.

References

1. Warburg O (1956) On the origin of cancer cells. Science 123:302–315
2. Vaupel P, Hockel M (2000) Blood supply, oxygenation status and metabolic micromilieu of breast cancers: characterization and therapeutic relevance. Int J Oncol 17:869–879
3. Wouters BG, Weppler SA, Koritzinsky M, Landuyt W, Nuyts S, Theys J, Chiu RK, Lambin P (2002) Hypoxia as a target for combined modality treatment. Eur J Cancer 38:240–257
4. Hasper D, Hummel M, Kleber FX, Reindl I, Volk HD (1998) Systemic inflammation in patients with heart failure. Eur Heart J 19:761–765
5. Gray LH, Conger AD, Ebert M, Hornsey S, Scott OC (1953) Concentration of oxygen dissolved in tissue at time of irradiation as a factor in radiotherapy. Br J Radiol 26:638–648
6. Rofstad EK (2000) Microenvironment-induced cancer metastasis. Int J Radiat Biol 76:589–605
7. Canning MT, Postovic LM, Clarke SH, Graham CH (2001) Oxygen mediated regulation of gelatinase and tissue inhibitor of matalloproteinase 1 expression by invasive cells. Exp Cell Res 367:88–94

8. Hasan NM, Adams GE, Joiner MC, Marshall JF, Hart IR (1998) Hypoxia facilitates tumor cell detachment by reducing expression of surface adhesion molecules and adhesion to extracellular matrices without loss of cell viability. Br J Cancer 77:1799–1805
9. Menger MD, Plusczyk T, Vollmar B (2001) Microcirculatory derengements in acute pancreatitis. J Hepatobiliary Pancreatic Surg 8:187–194
10. Saliba E, Henrot A (2001) Inflammatory mediators and neonatal brain damage. Biol Neonate 79:224–227
11. Stanimirovic D, Satoh K (2000) Inflammatory mediators of cerebral endothelium: a role in ischemic brain inflammation. Brain Pathol 10:113–126
12. Polosukhin VV (2000) Ultrastructure of the blood and lymphatic capillaries of the respiratory tissue during inflammation and endobronchial laser therapy. Ultrastruct Pathol 24:183–189
13. Hierholzer C, Harbrecht BG, Billiar TR, Tweardy DJ (2001) Hypoxia-inducible factor-1 activation and cyclo-oxygenase-2 induction are early reperfusion-independent inflammatory events in hemorrhagic shock. Arch Orthop Trauma Surg 121:219–222
14. De la Monte SM, Neely TR, Cannon J, Wands JR (2000) Oxidative stress and hypoxia-like injury cause Alzheimer-type molecular abnormalities in central nervous system neurons. Cell Mol Life Sci 57:1471–1481
15. Paulus WJ (1999) How are cytokines activated in heart failure? Eur J Heart Failure 1:309–312
16. Seta Y, Shan K, Bozkurt B, Oral H, Man DL (1996) Basic mechanism in heart failure: the cytokine hypothesis. J Cardiac Failure 2:243–249
17. Hayward CS, Macdonald PS, Keogh AM (2001) Inhaled nitric oxide in cardiology. Expert Opin Investig Drugs 10:1947–1956
18. Das UN (2000) Free radicals, cytokines and nitric oxide in cardiac failure and myocardial infarction. Mol Cell Biochem 215:145–152
19. Vanderheyden M, Kersshot E, Paulus WJ (1998) Proinflammatory cytokines and endothelium-dependant vasodilatation in the forearm: serial assessment in patients with congestive heart failure. Eur Heart J 19:747–752
20. Ostadal B, Urbanova D, Ressl J, Prochazka J, Pelouch V, Widimsky J (1981) Changes of the right and left ventricles in rats exposed to intermittent high altitude hypoxia. Cor Vasa 23:111–120
21. Turner L, Scotton C, Negus R, Balkwill F (1999) Hypoxia inhibits macrophage migration. Eur J Immunol 29:2280–2287
22. Takahashi M, Nishihira J, Shimpo M, Mizue Y, Ueno S, Mano H, Kobayashi E, Ikeda U, Shimada K (2001) Macrophage migration inhibitory factor as a redox-sensitive cytokine in cardiac myocytes. Cardiovasc Res 52:438–445
23. Naldini A, Carraro F, Silvestri S, Bocci V (1997) Hypoxia affects cytokine production and proliferative responses by human peripheral mononuclear cells. J Cell Physiol 173:335–42
24. Le YJ, Corry PM (1999) Hypoxia-induced bFGF gene expression is mediated through the JNK signal transduction pathway. Mol Cell Biochem 202 (1–2): 1–8

25. Tarnawski A, Szabo IL, Husain SS, Soreghan B (2001) Regeneration of gastric mucosa during ulcer healing is triggered by growth factors and signal transduction pathways. J Physiol Paris 95:337–344
26. Corpechot C, Barbu V, Wendum D, Kinnman N, Rey C, Poupon R, Housset C, Rosmorduc O (2002) Hypoxia-induced VEGF and collagen I expressions are associated with angiogenesis and fibrogenesis in experimental cirrhosis. Hepatology 35:1010–1021
27. Elson DA, Thurston G, Huang LE, Ginzinger DG, McDonald DM, Johnson RS, Arbeit JM (2001) Induction of hypervascularity without leakage or inflammation in transgenic mice overexpressing hypoxia-inducible factor-1alpha. Genes Dev 15:2520–2532
28. Falanga V, Zhou L, Yufit T (2002) Low oxygen tension stimulates collagen synthesis and COL1A1 transcription through the action of TGF-beta1. J Cell Physiol 191:42–50
29. Tajima R, Kawaguchi N, Horino Y, Takahashi Y, Inou K, Torii S, Kitagawa Y (2001) Hypoxic enhancement of type IV collagen secretion accelerates adipose conversion of 3T3-L1 fibroblasts. Bioch and Biophys Acta 1540:179–187
30. Fedele AO, Whitelaw ML, Peet DJ (2002) Regulation of gene expression by the hypoxia-inducible factors. Molecular interventions 2:229–243
31. Kivirriko K, Myllyharju J (1998) Prolyl-4-hydroxylatase and their protein disulfide isomerase unit. Matrix Biol 16:357–368
32. Takahashi Y, Takahashi S, Shiga Y, Yoshimi T, Miura T (2000) Hypoxic induction of prolyl 4-hydroxylase alpha (I) in cultured cells. J Biol Chem 275: 14139–14146
33. Huang LE, Gu J, Schau M, Bunn HF (1998) Regulation of hypoxia inducible factor 1a is mediated by an O2-dependant degradation domain via the ubiquitin proteasome pathway. Proc Nat Acad USA 95:7987–7992
34. Albina JE, Mastrofrancesco B, Vessella JA, Louis CA, Henry WL Jr, Reichner JS (2001) HIF-1 expression in healing wounds: HIF-1alpha induction in primary inflammatory cells by TNF-alpha. Am J Physiol Cell Physiol 281:C1971–C1977
35. Hollander AP, Corke KP, Freemont AJ, Lewis CE (2001) Expression of hypoxia-inducible factor 1alpha by macrophages in the rheumatoid synovium: implications for targeting of therapeutic genes to the inflamed joint. Arthritis Rheum 44:1540–1544
36. Hes FJ, van der Luijt RB, Lips CJ (2001) Clinical management of Von Hippel-Lindau (VHL) disease. Neth J Med 59:225–234
37. Kondo K, Kaelin WG Jr (2001) The von Hippel-Lindau tumor suppressor gene. Exp Cell Res 264:117–125
38. Richard S, Campello C, Taillandier L, Parker F, Resche F (1998) Haemangioblastoma of the central nervous system in von Hippel-Lindau disease. French VHL Study Group. J Intern Med 243:547–553
39. Oh M, Yauch RL, Lonergan KM, Whaley JM, Stemmer-Rachamimov AO, Louis DN, Gavin BJ, Kley N, Kaelin WG Jr, Iliopoulos O (1998) The von Hippel-Lin-

dau tumor suppressor protein is required for proper assembly of an extracellular fibronectin matrix. Mol Cell 1:959–968

40. Wykoff CC, Pugh CW, Harris AL, Maxwell PH, Ratcliffe PJ (2001) The HIF pathway: implications for patterns of gene expression in cancer. Novartis Found Symp 240:212–225
41. Reynolds TY, Rockwell S, Glazer PM (2002) Genetic instability induced by tumor hypoxia. Cancer Research 56:5754–5757
42. Prabhakar NR, Shenoy BC, Simonson MS, Cherniack NS (1995) Cell selective induction and transcriptional activation of immediate early genes by hypoxia. Brain Res 697:266–270
43. Ausserer WA, Bourrat-Floeck B, Green CJ, Laderoute KR, Sutherland RM (1994) Regulation of c-jun expression during hypoxic and low-glucose stress. Mol Cell Biol 14:5032–5042
44. Chandel NS, Vander Heiden MG, Thompson CB, Schumacker PT (2000) Redox regulation of p53 during hypoxia. Oncogene 19:3840–3848
45. Lakin ND, Jackson SP (1999) Regulation of p53 in response to DNA damage. Oncogene 18:7644–7655
46. Suzuki H, Tomida A, Tsuruo T (2001) Dephosphorylated hypoxia-inducible factor 1alpha as a mediator of p53-dependent apoptosis during hypoxia. Oncogene 20:5779–5788
47. Meek DW (1997) Post-translational modification of p53 and the integration of stress signals. Pathol Biol (Paris) 45:804–814
48. Gammack D, Byrne HM, Lewis CE (2001) Estimating the selective advantage of mutant p53 tumour cells to repeated rounds of hypoxia. Bull Math Biol 63:135–166
49. Portwine C (2000) P-53 the link between inflammation and cancer. Ped Res 47:573–582
50. Chan DA, Sutphin PD, Denko NC, Giaccia AJ. Role of propylhydroxylation in oncogenically stabilized hypoxia-inductible factor. J Biol Chem (in press)
51. Brune B, von Knethen A, Sandau KB (2001) Transcription factors P53 and HIT-1alpha as targets of nitric oxide. Cell Signal 13:525–533
52. Haddad JJ (2002) Recombinant human interleukin (IL)-1beta-mediated regulation of hypoxia-inducible factor-1alpha (HIF-1alpha) stabilization, nuclear translocation and activation requires an antioxidant/reactive oxygen species (ROS)-sensitive mechanism. European Cytokine Network 13:50–60
53. Laughner E, Taghavi P, Chiles K, Mahon PC, Semenza GL (2001) HER2 (neu) signaling increases the rate of hypoxia-inducible factor 1alpha (HIF-1alpha) synthesis: novel mechanism for HIF-1-mediated vascular endothelial growth factor expression. Mol Cell Biol 21:3995–4004

Fibrosis Perpetuates Hypoxia

CHAPTER 7

Whether it is untreated pneumonia, an unremoved foreign body like a splinter, or cardiac insufficiency resistant to treatment, all chronic inflammation progresses toward fibrosis. Hepatitis is another example: either inflammation disappears or the disease become chronic and evolves toward cirrhosis.One of the consequences of hypoxia is the secretion of extracellular matrix, particularly of collagen. These deposits perpetuate hypoxia, alter tissue architecture, and further damage the genome.

Fibrosis is a consequence of nonresolutive inflammation (1, 2). All fibrotic lesions, such as those in the liver (cirrhosis), joints (polyarthritis), and lung (idiopathic lung fibrosis) or radiation-induced fibrosis, whatever the type of stress which initiated them, display common features (1). In the early phase, there is tissue inflammation with lymphocytes and macrophages, angiogenesis and abundant myofibroblasts. These fibroblasts deposit extracellular matrix, which is abnormal both in quantity and quality. In every case, hypoxia appears to precede tissue changes. It is beyond the scope of this chapter to review them all, and therefore will only give a few examples.

Keloids Are Hypoxic

The healing of a deep surface wound begins with the formation of granulation tissue, initially in an inflammatory milieu. The sequel may be a hypertrophied scar or keloid. Hypertrophied scars and keloids are hypoxic because of microvascular occlusion. Hypoxia stimulates excessive production of collagen, which forms the bulk of these lesions, from fibroblasts and myofibroblasts (3).

Ethanol-Induced Liver Cirrhosis Is Hypoxic

The hepatotoxicity of alcohol results from the metabolic disturbances associated with the oxidation of ethanol via the liver alcohol dehydrogenase (ADH) pathway and the redox changes produced by the generated NADH, which in turn affect the metabolism of lipids, carbohydrates, proteins and purines (4). In addition to ADH, ethanol can be oxidized by liver microsomes. Induction of this microsomal pathway contributes to increased acetaldehyde generation, with formation of protein adducts, resulting in antibody production and enzyme inactivation; it is also associated with a striking impairment of the capacity of the liver to utilize oxygen. Moreover, acetaldehyde promotes glutathione depletion, free radical-mediated toxicity and lipid peroxidation. Ethanol and its metabolite acetaldehyde increase collagen accumulation (4, 5).

Exaggeration of the redox change by relative hypoxia, which prevails physiologically in the perivenular zone, contributes to the exacerbation of ethanol-induced lesions (4).

Radiation-Induced Fibrosis Is Hypoxic

Radiation-induced fibrosis is a relatively uncommon complication of high-dose radiation therapy or accidental overexposure. It has been described in vivo in several tissues, including the skin, lung, heart, and liver (1). It develops months or years following excessive irradiation. It usually starts with local inflammation in the high-dose target area and evolves slowly toward fibrosis or even necrosis.

Moderate hypoxia is detected shortly after irradiation, before the onset of functional or histopathologic changes. There is a striking correlation between the severity of hypoxia and the subsequent development of fibrosis (6, 7).

Hypoxia Increases Extracellular Matrix Deposition

There is clinical evidence that hypoxia is enough to increase the secretion of extracellular matrix. It has been reported that cardiac ligature or exposure of a transplanted tendon to anoxia results in transient chondrogenesis followed by enchondral ossification (8). Similarly, Busher (9) reports the case of a patient who underwent gastric tube reconstruction. Following cardiac arrhythmia, he developed hypoxia and ischemic necrosis over 5 cm of the proximal gastric tube. Three weeks later that area was ossified; trabecular bone was present along the entire length of the constricted gastric tube.

Ectopic ossification has been associated with several conditions in both neoplastic and non-neoplastic tissue (9–12). Paraplegic patients frequently develop ectopic ossification (9, 13). But there is no significant difference in ectopic bone formation between paraplegic rabbits and nonparaplegic rabbits under the same conditions of immobilization and passive movement of the posterior legs. For Izumi (13), the reason for ectopic bone formation may simply be poor oxygenation because of blood stasis.

Calcium mineral deposition in the atherosclerotic plaque also results from hypoxia. The vascular cell, when exposed to hydrogen peroxide, differentiates into an osteoblastic cell (14). This effect is counteracted by antioxidants. Oxidative stress modulates the differentiation of bone and vascular cells in opposite ways (14).

To say that hypoxia is responsible for the formation of cartilage is to acknowledge the link between biochemistry and physics. A fibroblast is a mobile cell as witnessed by its differentiation into a macrophage. To subject this fibroblast to hypoxia is to transform it into a static, cartilaginous cell embedded in its extracellular matrix. If you expose it to further shear stress, you relegate it to a bony prison cell.

Extracellular Matrix Deposition Mimics Hypoxia

Collagen is a vital component of the extracellular matrix of both muscles and tendons (15). It acts as a scaffold to maintain muscle shape and allows seven distribution of force and plays a crucial role in the mechanical properties of the tendons.

Under normal circumstances, collagen is continually being synthesized and degraded throughout life. Increased mechanical stress, which causes muscle hypertrophy, stimulates collagen synthesis. The concentration of other extracellular molecules (laminin, fibronectin, and vitronectin) is also modulated by mechanical stress (16).

Oxygen depletion seems to have the same effect on cell differentiation as exposure to mechanical constraints. The extracellular matrix and\or low oxygen tension differentiates stem cells into chondrocytes (19). Cartilage is practically avascular; nutrients are transported from the synovial fluid and vascularized subchondral bone (20). Oxygen tensions within cartilage are therefore significantly lower than in vascularized tissues ranging from 2.7% to 7.5% oxygen (20–22). Low oxygen tension, as well as shear stress, promotes and induces redifferentiation of dedifferentiated bovine articular chondrocytes (24).

It is beyond the scope of this book to elucidate the complex link between extracellular matrix deposition and hypoxia. What is clear is that extracellular matrix deposition induces the same pathway as hypoxia. A stretch of cardiomyocytes cultured on silicone membranes activates extracellular matrix deposition. There are similar protein kinase cascades of phosphorylation and a similar increase in protein synthesis, and in the expression of genes such as c-*fos*, c-*myc*, and c-*jun* (27), as when there is exposure to hypoxia. Similarly, mechanical stress activates HIF. This activation of HIF results, as always, in the release of "growth factors" like VEGF (28).

Extracellular Matrix Deposition Perpetuates Hypoxia

As a gas, oxygen diffuses poorly, twenty times less than CO_2. One example comes from Alsace. In the torrents of the Vosges, trout is abundant. A few kilometers away, on the plain of the region, the resurgence of ground waters has flooded old gravel pits. Fishermen pay high prices for their permits and fish for tench, carp, and pikes. But they never find trout in these waters. This fish needs high concentrations of oxygen to survive. The lively waters of the torrents provide it, but the stagnant waters of the lakes contain less oxygen. Since trout is a highly valued fish, fishermen have designed air injectors to add oxygen to the water. A pipe connected to the ventilation system blows air into the lake. Trout can then live in these still, oxygen-enriched waters.

In the human body, as in the Alsatian lakes, oxygen diffuses poorly. Slight changes in distances and chemical content are enough to change markedly the amount of oxygen available to the cell. Poor oxygen distribution because of extracellular matrix deposition is confirmed by pathology. Even fairly mild thickening of the alveolar walls (pulmonary edema, acute respiratory distress syndrome, interstitial fibrosis, lymphangitic carcinomatosis) renders them relatively impermeable to oxygen ("diffusion barrier") (25, 26). During inflammation and subsequent fibrosis there is accumulation of extracellular matrix. This diffusion barrier results in chronic cellular hypoxia.

Conclusion

Anaerobic glycolysis is responsible for vascular dilatation, release of polypeptides, and increased secretion of collagen, which together constitute inflammation. If hypoxia is not corrected quickly by removing the foreign body, regulating blood flow or treating infection, fibrosis will set in. Extracellular matrix deposits will perpetuate hypoxia. This inevitable consequence of nonresolving inflammation, fibrosis, will in turn present all the signs of hypoxia: neovascularization, extracellular matrix deposition, release of growth factors and cytokines.

References

1. Schwartz L, Balosso J , Baillet F, Brun B, Amman JP, Sasco A (2002) Cancer, the role of extra-cellular disease: a hypothesis. Med Hypotheses 58:340–346
2. Balkwill F, Mantovani A (2001) Inflammation and cancer: back to Virchow? Lancet 357:539–545
3. Kischer CW (1992) The microvessels in hypertrophic scars, keloids and related lesions: a review. J Submicrosc Cytol Pathol 24:281–296
4. Lieber CS (1993) Aetiology and pathogenesis of alcoholic liver disease. Baillieres Clin Gastroenterol 7:581–608
5. Bardag-Gorce F, French BA, Li J, Riley NE, Yuan QX, Valinluck V, Fu P, Ingelman-Sundberg M, Yoon S, French SW (2002) The importance of cycling of blood alcohol levels in the pathogenesis of experimental alcoholic liver disease in rats. Gastroenterology 123:325–335
6. Franko AJ, Sharplin J (1994) Development of fibrosis after lung irradiation in relation to inflammation and lung function in a mouse strain prone to fibrosis. Radiat Res 140:347–355
7. Vujaskovic Z, Anscher MS, Feng Q, Rabbani ZN, Amin K, Samulski T, Dewhirst MW, Haroon ZA (2001) Radiation-induced hypoxia may perpetuate late normal tissue injury. Int J Radiat Biol Phys 50:851–855
8. Danis A (1992) Soft tissue ossification: mechanism. Bull Mem Acad R Belg 147:298–306
9. Buscher HC, Van Lanschot JJ, Mulder AH, Tilanus HW (1995) Heterotopic ossification induced by hypoxia in a retrosternal gastric tube following transhiatal oesophagectomy. J Clin Pathol 48:177–178
10. Ackerman NB (1968) Bone formation in the gastric mucosa following subtotal gastrectomy in rats. Exp Mol Pathol 9:125–130

11. An HS, Ebrahem N, Kim K, Jackson WT, Kane JT (1987) Heterotopic ossification and pseudoarthrosis in the shoulder following encephalitis. Clin Orthop 219:291–298
12. Asa DK, Bertorini TE, Pinals RS (1986) Myositis ossificans circumscripta: a complication of tetanus. Am J Med Sci 292:40–43
13. Izumi K (1983) Study of ectopic bone formation in experimental spinal cord injured rabbits. Paraplegia 21:351–363
14. Mody N, Parhami F, Sarafian TA, Demer LL (2001) Oxydative stress modulates differentiation of vascular and bone cells. Free Rad Biol Med 31:509–519
15. Bishop JE. Regulation of cardiovascular collagen deposition by mechanical forces. Mol. Med. Today 1998;4:69–75
16. Loeser RF (2000) Chondrocyte integrin expression and function. Biorheology 37:109–116
17. Wilson DJ (1986) Development of avascularity during cartilage differentiation in the embryonic limb. differentiation 30:183–187
18. Jin M, Frank EH, Quinn TM,Hunziker EB,Grodzinsky AJ (2001) Tissue shear deformation stimulates proteoglycan and protein biosynthesis in bovine cartilage explants. Arch Bioch Biop 395:41–48
19. Von Der Mark K, Conrad G (1979) Cartilage and cell differentiation:a review. Clin Orthop 139:185–205
20. Murphy C, Sambanis A (2001) Effect of oxygen tension on chondrocyte extracellular matrix accumulation. Connective Tissue Res 42:87–96
21. Brighton CT, Heppenstall RB (1971) Oxygen tension in zone of the epiphyseal plate, the metaphysis and the diaphysis. Am J Bone Joint Surg 53:719–728
22. Ellis S, Velayutham M, Velan S, Petersen E, Zweiler J, Kuppusamy P, Spencer R (2001) EPR oxygen mapping of engineered cartilage in reactor grown in a hollow-fiber bioreactor. Mag Res Med 46:819–826
23. Butler AJ, Eagleton MJ, Wang D, Howell RL, Strauch AR, Khasgiwala V, Smith HC (1991) Induction of the proliferative phenotype in differentiated myogenic cells by hypoxia. J Biol Chem 266:18 250–18 258
24. Domm C, Schunke M, Christesen K, Kurz B (2002) Redifferentiation of dedifferentiated bovine articular chondrocytes in alginate culture under low oxygen tension osteoarthritis. Cartilage 10:13–22
25. Borovetz HS, Chen CC, Hardesty RL (1981) Numerical simulation of the transient transport of inert gases in lung tissue. Phys Med Biol 26:401–411
26. Cherniack RM (1992) Evaluation of respiratory function in health and disease. Dis Mon 38:505–576
27. Yamazaki T, Komuro I, Yazaki Y (1995) Molecular mechanism of cardiac cellular hypertrophy by mechanical stress. J Mol Cell Cardiol 27:133–140
28. Kim CH, Cho YS, Chun YS, Park JW, Kim MS (2002) Early expression of myocardial HIF-1 alpha in response to mechanical stresses: regulation by stretch-activated channels and the phosphatidylinositol 3-kinase signaling pathway. Circ Res 90:25–33

Metastasis: The Hunt for Food

CHAPTER 8

In order to form metastases, cells have to escape, devour the matrix, which limits their growth, invade the stroma, infiltrate blood and lymphatic vessels, survive circulatory shear stress and, finally, be able to adhere to the endothelial cells of the organ that they will colonize.
The ability of cancer cells to metastasize appears as a direct consequence of hypoxia. The decreased concentration in oxygen is responsible for a loosening of attachment of the cell to the mesenchyme and the induction of proteolytic enzymes. These cells are able to break cell–matrix and cell–cell attachments and to travel to distant sites. Metastatic dissemination is foreseeable. Tumor cells travel to hypoxic sites where there is a release of by-products of anaerobic glycolysis.

When we were children in the Alsace, we used to spread nets and put a small ring around the foot of the migratory birds. Scientists could create maps of their migrations. Each year at the same time, birds left and joined other birds at certain latitudes. The only way to prevent these migrations is to offer the birds modern comforts. In order to keep storks from leaving in the winter (they are a tourist attraction), zoos feed them and heat their surroundings.

Stork migration resembles that of cancer cells: reproducible and responding to external perturbation.

Extracellular stress like low oxygen tension in the primary tumor is associated with metastasis in soft tissue sarcoma, cervical carcinoma, and carcinoma of the head and neck (2, 3). Low glucose concentration, high lactate concentration and low extracellular pH may induce metastasis by mechanisms similar to hypoxia (4).

Hypoxia Induces Genes Involved in the Metastatic Cascade

Multiple mechanisms are involved in hypoxia-induced metastasis (4). Hypoxia induces genes involved in the metastatic cascade-like matrix metalloproteinases (MMPs), which degrade the surrounding stroma and decrease the secretion of its inhibitors (6, 7). Urokinase (u-PA) is another protease, a proteolytic endogenous activator of the thrombolytic mediator plasminogen; its secretion is induced by hypoxia in cancer cells, resulting in increased invasiveness and tumor progression.

Similarly, cell surface integrins and other adhesion molecules are down-regulated by hypoxia, decreasing cell-to-cell adhesion (7).

The tumor environment is also more hydrated and therefore softer and easier to invade than the normal organ. Hyaluronan, a high-molecular-weight glycosaminoglycan of the extracellular matrix, hydrates the stroma, providing an environment that facilitates cell movement, thereby participating in the process of cancer invasion and metastasis (8). The stromal secretion of hyaluronan appears to be a consequence of high lactate secretion by cancer cells.

Furthermore, as seen in previous chapters, tumor cells exposed to hypoxia have been shown to up-regulate the expression of vascular endothelial growth factor (VEGF), inducing neoangiogenesis, thus stimulating tumor cell migration (9,10).

Metastatic Dissemination Is Foreseeable

Just as cell differentiation can indicate the origin of cancer, its metastatic progression depends on the site of the primary tumor. Choroidal melanomas metastasize almost exclusively to the liver; squamous cell carcinoma of the eyelid to the draining lymph nodes. Breast cancer travels to the axillary lymph nodes, the liver, the lungs, the brain, but not the heart or the spleen. Adrenal gland metastases indicate primary lung cancer.

It is still unclear why particular cancers preferentially metastasize to certain sites. The possibilities usually discussed involve differential survival and proliferation at these sites or selective trapping with or without preferential homing (11).

Metastases Travel to Hypoxic Sites

Metastases carried by the blood can travel only to vascularized areas and can only take hold if cancer cells have the adequate anchorage mechanisms. But this is only one part of the picture. The most oxygenated sites, the ones with the highest vascularity, are not the ones where metastases most often appear. Thus, metastases to the kidney, spleen, heart, or muscle are rare. Metastases are more frequent in hypoxic areas such as the lymph nodes, the area of the brain at the junction of the white and gray matter, bone, and adrenal glands.

Throughout evolution, both prokaryotic and eukaryotic cells have developed a variety of biochemical mechanisms to define the direction and proximity of extracellular stimuli. This process is essential for the cell to reply properly to the environmental cues that determine cell migration, proliferation, and differentiation. Chemotaxis is the cellular response to chemical attractants that direct cell migration, a process that plays a central role in many physiological situations, such as host immune responses, angiogenesis, wound healing, embryogenesis, and neuronal patterning among others (12). Cellular migration follows proteic gradients (chemotaxis). Just like bacteria, human cells travel toward nutritional sources.

Numerous in vivo methodologies have documented the invasive behavior of cancer cells through normal parenchyma. Cancer cell locomotion has been assessed with a number of in vitro assays including the Boyden chamber and other chemotaxis assays, colloidal gold cell tracking, analysis of migration of tumor cells from spheroids, confronting cultures of cancer cells with aggregates of non-neoplastic tissue, time-lapse video microscopy, electron microscopic examination of the cytomorphologic correlates of cell motility, the radial dish assay, and quantitative enzyme immunoassay of proteins associated with invasion (e.g., laminin).

All these assays point out a simple fact: cancer cells move toward food. Cytokines and growth factors such as epidermal growth factor (EGF), basic fibroblast growth factor (bFGF), platelet-derived growth factor (PDGF), nerve growth factor (NGF), interleukin 2 (IL-2), transforming growth factors alpha, and tumor necrosis factor-alpha (TNF-alpha) enhance tumor cell motility (13, 14).

Polypeptides, such as growth factors and cytokines, are sent out into the circulation by hypoxic tissues after anaerobic glycolysis, which is partial combustion. The cancer cells move toward the following gradients of growth factors and cytokines. These growth factors are in reality polypeptides captured, metabolized, and digested by the cells. It is exactly as if the cells move toward a nurture-rich environment.

Cardiac Infarction: Confirmation of the Role of Hypoxia in Cell Migration

Hypoxia is thought to provide physiological pressure in tumors selecting for metastatic cell phenotypes (4). More likely, cell migration is a simple consequence of hypoxia. Hypoxia decreases cell-to-cell attachment, increases protease activity, and promotes cell migration

(7). In cases of angina pectoris or infarction there is a reduction in oxygen supply. Protease activity is increased. For example, in a model of cardiac infarction, myocytes show an increase of MMP-3 and MMP-9 expression and a concomitant decrease of TIMP-1 (15). Similarly, urokinase activity is increased. There is concomitant release of chemotactic cytokines and growth factors. Cardiac myocytes exposed to anoxia–reoxygenation generate a chemotactic gradient for polymorphonuclear neutrophil (PMN). Supernatants obtained from neonatal mouse hypoxic cardiac myocytes promote PMN migration across mouse myocardial endothelial cell monolayers (16).

References

1. Orr FW, Wang HH, Lafrenie RM, Scherbarth S, Nance DM (2000) Interactions between cancer cells and the endothelium in metastasis. J Pathol 190:310–329
2. Vaupel P, Kelleher DK, Hockel M (2001) Oxygen status of malignant tumors: pathogenesis of hypoxia and significance for tumor therapy. Semin Oncol 28:29–35
3. Hockel S, Schlenger K, Vaupel P, Hockel M (2001) Association between host tissue vascularity and the prognostically relevant tumor vascularity in human cervical cancer. Int J Oncol 19:827–832
4. Rofstad EK (2000) Microenvironment-induced cancer metastasis. Int J Radiat Biol 76:589–605
5. Del Rosso M, Fibbi G, Pucci M, D'Alessio S, Del Rosso A, Magnelli L, Chiarugi V (2002) Multiple pathways of cell invasion are regulated by multiple families of serine proteases. Clin Exp Metastasis 19:193–207
6. Hasan NM, Adams GE, Joiner MC, Marshall JF, Hart IR (1998) Hypoxia facilitates tumor cell detachment by reducing expression of surface adhesion molecules and adhesion to extracellular matrices without loss of cell viability. Br J Cancer 77:1799–1805
7. Canning MT, Postovic LM,Clarke SH,Graham CH (2001) Oxygen mediated regulation of gelatinase and tissue inhibitor of matalloproteinase 1 expression by invasive cells. Exp Cell Res 367:88–94
8. Stern R, Shuster S, Neudecker BA, Formby B (2002) Lactate stimulates fibroblast expression of hyaluronan and CD44: the Warburg effect revisited. Exp Cell Res 276:24–31
9. Rofstad EK, Danielsen T (1999) Hypoxia-induced metastasis of human melanoma cells: involvement of vascular endothelial growth factor-mediated angiogenesis. Br J Cancer 80:1697–1707

10. Plasswilm L, Tannapfel A, Cordes N, Demir R, Hoper K, Bauer J, Hoper J (2000) Hypoxia-induced tumour cell migration in an *in vivo* chicken model. Pathobiology 68:99–105
11. Moore MA (2001) The role of chemoattraction in cancer metastases. Bioessays 23:674–676
12. Manes S, Mira E, Gomez-Mouton C, Lacalle RA, Martinez C (2000) Cells on the move: a dialogue between polarization and motility. IUBMB Life 49: 89–96
13. Chicoine MR, Silbergeld DL (1997) Mitogens as motogens. J Neurooncol 35: 249–257
14. Hitchon C, Wong K, Ma G, Reed J, Lyttle D, El-Gabalawy H (2002) Hypoxia-induced production of stromal cell-derived factor 1 (CXCL12) and vascular endothelial growth factor by synovial fibroblasts. Arthritis Rheum 46:2587–2597
15. Romanic AM, Burns-Kurtis CL, Gout B, Berrebi-Bertrand L, Ohstein EH (2001) Matrix metalloprotease expression in cardiac myocytes following myocardiac infarction in the rabbit. Life Sci 68:799–814
16. Rui T, Cepinskas G, Feng Q, Ho YS, Kvietys PR (2001) Cardiac myocytes exposed to anoxia-reoxygenation promote neutrophil transendothelial migration. Am J Physiol Heart Circ Physiol 281:H440–H447

Carcinogens Target Cell Respiration and Induce Glycolysis

CHAPTER 9

Genetic instability has been implicated prominently in tumor formation. The main evidence comes first from the discovery of chromosomal aberrations and then from mutations of oncogenes or tumor suppressor genes. These anomalies of the genome are usually not a primary event but secondary to stress of extracellular origin such as hypoxia or physical constraints. Carcinogens (either viral, chemical, or physical) target cell respiration and induce glycolysis. Thus carcinogens mimic hypoxia and induce mutations.
Inhibition of glycolysis markedly decreases carcinogenesis, thus confirming the key role of cell respiration.

In cases of tuberculosis, it can be difficult to isolate the Koch bacillus. Microscopic examination results of intubation liquid, or even of cultures, can be negative. But polymerase chain reaction (PCR) analysis of a biopsy specimen of the suspect zone will always find traces of the Koch bacillus. This is not the case for cancer.

Such a molecular signature for carcinogens does not exist in cancer. Smoking is responsible for most cases of lung cancer. But a precise analysis of the lung cancer genome does not reveal a constant anomaly, a precise signature. In 70% of lung carcinomas, there is a mutation of the p53 gene of the "*ras*" oncogene. Disturbances of the tumoral genome (deletions, mutations, and overexpression) are constant but not reproducible.

In experimental carcinogenesis, toxicologists identify two successive stages. The first is tumoral initiation, the second is promotion. An initial toxic substance, DMBA for example, is injected into the animal; in the following weeks, tumoral promotion is made possible by daily administration of TPA. When used alone, in the dosages already mentioned, neither DMBA nor TPA is a carcinogen.

This multistage model of carcinogenesis indicates that the first step in this process – the initiation stage – is irreversible. Mutations in the cell's genome are widely perceived to explain this irreversibility.

The second step in carcinogenesis – the promotion stage – appears to involve the clonal expansion of an initiated stem cell, which, because it is unable to terminally differentiate, accumulates as a focus of nonterminally differentiated cells. Examples of such foci might be papillomas of the skin, enzyme-altered foci of the liver, polyps of the colon, and nodules of the breast. Obviously, this process must require stimulation of cell division (i.e., it must be mitogenic), at least with respect to the initiated cell. As demonstrated in experimental animals, this stage is potentially interruptible and reversible.

Are Mutations the Cause or the Consequence of Carcinogenesis?

The multihit process of carcinogenesis is thought to involve a small number of aberrant genetic events culminating in malignant transformation (1, 2). This theory predicts a few DNA lesions as explanations of carcinogenesis (1, 2).

The discovery and subsequent improvement of PCR techniques have resulted in rapid and relatively easy analysis of the tumoral cell genome. Over the past 25 years, cancer researchers have enumerated a bewildering array of molecular alterations (most of them mutations) associated with cellular malignancy (3). Cancer cells harbor as many as one hundred thousand mutations, a much greater number than that predicted by the cellular theory of cancer.

These mutations are not specific to cancer and often occur in ordinary aging. The organs affected are not those that most commonly become cancerous.

For example, a plasmid with no particular function (lac Z) is inserted in a line of transgenic mice, with no mutations being foreseen. Analysis of this plasmid shows the rate and nature of mutations. The spectrum of spontaneous point mutations is determined in the brain, heart, liver, spleen, and small intestine in young and old mice (4). While the small intestine has the highest spontaneous mutation rate of all tissues tested, tumors in this tissue occur at very low frequencies. As some authors have stated, "Other factors than somatic mutation rate alone play a role in determining susceptibility of an organ to tumor formation" (4).

Are these mutations the cause of cancer or the consequence of extracellular events?

Mutation as a Response to Extracellular Stress

Our cells, like bacteria, have to adapt to the biochemical and physical changes of a stressful and hostile environment. There is ample evidence that mutation is a means of adapting to these perturbations.

In bacteria, starving conditions, changes in pH, or changes in oxygen content induce mutagenesis (5, 6). Similarly, change in oxygen concentrations is mutagenic in the worm *Caenorhabditis elegans* (7).

Similarly, in mammalian cells, nongenotoxic stress, such as heat or serum starvation, acidosis or hypoxia, can induce a mutator phenotype with persistent pronounced genetic instability. Exposures to acidic and hypoxic environments have been shown to produce a wide range of cytogenetic changes. These changes include: (a) increases in mutation frequencies; (b) deficits in DNA repair; (c) DNA over-replication and gene amplification; (d) induction of chromosomal fragile sites, triggering genomic rearrangements; and (e) changes in gene expression (8, 9). Mutations in the tumor suppressor gene p53 may play an important role in regulating the adaptive response of tumor cells to hypoxia or acidosis by enhancing their survival (10, 11).

Similarly, mechanical stress is mutagenic. Chronic mechanical irritations induce mutations. Male rats were fed uracil, a component of RNA. This diet is responsible for multiple lithiasis of the urinary tract. As a consequence of the irritation of the urinary tract, the rats first developed papillomatosis and eventually bladder cancer. Compared with untreated animals, treated rats had a three- to fivefold increase in mutations in the bladder (12).

Heat Shock Proteins and Mutations

The heat shock response is a primitive and well-conserved cellular defense mechanism (13). One of the strongest and most noticeable responses of bacteria to a range of stress conditions, such as high temperature, low pH, osmolarity, substrate limitation, is the dramatic induction of a large number of general stress proteins (14, 15). These heat shock proteins (Hsp) play a crucial role in governing proper protein assembly, folding, and transport. They have a dual role in stress response and signal transduction. The synthesis of these stress-induced proteins promotes the survival of bacteria and their resistance to the action of humoral and cell-mediated protective factors of the host.

In mammals, Hsp are induced by stresses as diverse as heat, starvation, or injury (16). Stress, whatever its cause, appears to induce the synthesis of Hsp. Hypoxia also induces Hsp (17), as does mechanical stress. In isolated, erythrocyte-perfused rabbit heart there are increased levels of Hsp after a mild mechanical stress (18–20). Similarly, acute elevation in blood pressure results in Hsp expression in rat aortas.

As in invertebrates, Hsp protect the cells of mammals that are exposed to numerous types of injuries including heat, oxidative stress, or treatments with anticancerous and apoptosis-inducing agents (21, 22). In the cardiovascular system, this enhanced Hsp synthesis leads to a transient but powerful increase in tolerance to such endangering situations as ischemia, hypoxia, oxidative injury, and endotoxemia or simple exercise (23).

One potential mechanism involves the ability of the heat shock response to inhibit inflammatory responses, which as we saw in the previous chapter result from hypoxia. The heat shock response inhibits the expression of proinflammatory cytokines such as tumor necrosis factor and IL-1 beta (24).

In *Drosophila*, when the function of Hsp is impaired, there is an increased rate of mutations and multiple abnormal flies are created (25). During stress, Hsp become diverted from their function in signal transduction, through their affinity for denatured proteins. The heat shock protein Hsp90 supports diverse but specific signal transducers and lies at the interface of several developmental pathways. When *Drosophila* Hsp90 is mutant or pharmacologically impaired, a phenotypic variation affecting nearly any adult structure is produced, with specific variants depending on genetic background and occurring both in laboratory strains and in wild populations. Multiple, previously silent, genetic determinants produce these variants and when enriched by selection they rapidly became independent of the Hsp90 mutation. When Hsp90 buffering is compromised, for example by temperature, mutants are expressed and selection can lead to the continued expression of these traits, even when Hsp90 function is restored (25).

Carcinogens Mimic Extracellular Stress and Induce Hsp

Modification of DNA by carcinogens has long been recognized as an early event in carcinogenesis and many DNA adducts have been characterized. Disturbances of the tumoral genome (deletions, mutations, overexpression) are constant but not reproducible. However, most carcinogens do not react with DNA directly (26).

Therefore, it seems that the action of carcinogens on DNA is not directly responsible for the mutations often described in cancer. It is more likely that we are seeing the consequence of nonselective stress. These genetic disturbances are either the result of direct extracellular stresses or the result or their analogs: carcinogens (26).

Exposure of cells to carcinogens, directly mutagenic or not, most frequently leads to increased expression of Hsp. Hsp are induced by exposure to a carcinogen such as dimethylbenz[a]anthracene (27), bromobenzene, cadmium, cyclophosphamide(28), diethyl-nitrosamine (29), 2-acetylaminofluorene (30), or alcohol (31).

Carcinogens Target Cellular Respiration

In 1955, Warburg (32) stated: “Carcinogenesis starts with the irreversible injuring of respiration. The irreversible injuring of respiration is followed, as the second phase of cancer formation, by a long struggle by the injured cells to maintain their structure, in which a part of the cells perish from lack of energy, while another part succeed in replacing the irretrievably lost respiration energy by fermentation energy. Because of the morphological inferiority of fermentation energy, highly differentiated body cells are transformed into undifferentiated cells that grow wildly – cancer cells.

I may mention a few respiratory poisons. A strong, specific respiratory poison is arsenious acid which, as every clinician knows, may produce cancer. Hydrogen sulfide and many of its derivatives are also strong, specific respiratory poisons. We know today that certain hydrogensulfide derivatives, thiourea and thioacetamide, with which citrus fruit juices have been preserved in recent times, induce cancer of the liver and gall bladder in rats.

Urethane is a nonspecific respiratory poison. It inhibits respiration like a chemically indifferent narcotic, since it displaces metabolites from cell structures. In recent years it has been recognized that a subnarcotic dose of urethane causes lung cancer in mice in 100 percent of treatments. Urethane is particularly effective as a carcinogen, because in contrast to alcohol, it is not itself burned up on the respiring surfaces and, unlike ether or chloroform, it does not cytolize the cells. Any narcotic that has these properties may cause cancer upon chronic administration in small doses."

If the Rous virus (which expresses the *src* oncogene) is inoculated into the chorion of chick embryos, tumors originate in the course of a few days – as rapidly as the transplantation of cancer cells. Infection by this oncogenic virus results in rapidly increased glycolytic activity (33, 34).

Similarly, carcinogens inhibit mitochondrial respiration. For example, carcinogenic urethane, dimethylnitrosamine, 3-methylcholanthrene, benzo[a]pyrene, 7,12-dimethylbenz[a]anthracene, and aflatoxin B1 enhance glycolysis. Noncarcinogenic phenylurethane, ethylformate, chrysene, perylene, and pyrene have no effect (35). Similarly, other known carcinogens such as radiation, hormones, cytotoxic chemotherapy, or asbestos alter mitochondrial respiration and induce glycolysis (35–38).

Modern molecular biology confirms, albeit in different terms, the importance of fermentation during carcinogenesis. The aryl hydrocarbon (Ah) has occupied the attention of toxicologists for close to three decades. The carcinogenicity of well-known molecules like polycyclic aromatic hydrocarbons or halogenated dioxins appears to be mediated by the binding to this nuclear receptor (39). In the presence of a carcinogen, this receptor cannot bind to the aryl hydrocarbon nuclear transporter (ARNT). ARNT is released and in turn activates HIF, inducing glycolytic enzymes (39).

References

1. Volpe JP (1992) Can a non-tumorigenic dose of carcinogen cause two or more errors in a cell? Cancer Lett 64:91–97
2. Field JK (1992) Oncogenes and tumour-suppressor genes in squamous cell carcinoma of the head and neck. Eur J Cancer B Oral Oncol 28B:67–76
3. Hahn WC,Weinberg RA (2002) Mechanism of diseases:Rules for making human tumor cells. N E J M 347:1593–1603
4. Dollé ME, Snyder W, Dunson DB, Vijg J (2002) Mutational fingerprints of aging. Nucleic Acids Res 30:545–549
5. McKenzie GJ, Harris RS, Lee PL, Rosenberg SM. The SOS response regulates adaptive mutation. Proc Natl Acad Sci USA 97:6646–6651
6. Khil PP, Camerini-Otero RD (2002) Over 1000 genes are involved in the DNA damage response of Escherichia coli. Mol Microbiol 44:89–105
7. Honda Y, Honda S (2002) Oxidative stress and life span determination in the nematode Caenorhabditis elegans. Ann N Y Acad Sci 959:466–474
8. Reynolds TY, Rockwell S, Glazer PM. Genetic instability induced by the tumor microenvironment. Cancer Research 56:5754–5757
9. Rockwell S, Yuan J, Peretz S, Glazer PM (2001) Genomic instability in cancer. Novartis Found Symp 240:133–142
10. Pouyssegur J, Franchi A, Pages G (2001) pHi, aerobic glycolysis and vascular endothelial growth factor in tumour growth. Novartis Found Symp 240: 186–196
11. Royds JA, Dower SK, Qwarnstrom EE, Lewis CE (1998) Response of tumour cells to hypoxia: role of p53 and NFkB. Mol Pathol 51:55–61
12. Takahashi S, Ikeda Y, Okochi E, Cui L, Nagao M, Shirai T (2000) Mutation induction by mechanical irritation caused by uracil induced urolithiasis in Big Blue rats. Mutat Res 447:275–280
13. Xu Q, Schett G, Li C, Hu Y, Wick G (2000) Mechanical stress-induced heat shock protein 70 expression in vascular smooth muscle cells is regulated by Rac and Ras small G proteins but not mitogen-activated protein kinases. Circ Res 86:1122–1128
14. Basnakian IA, Bondarenko VM, Melnikova VA, Beliavskaia VA (2001) Stress-inducible bacterial proteins and virulence Zh Mikrobiol Epidemiol Immunobiol 5:101–108
15. Hecker M, Volker U (2001) General stress response of Bacillus subtilis and other bacteria. Adv Microb Physiol 44:35–91

16. Rickley TM, Belknap WR (1991) Comparison of the expression of several stress responsive genes in potato tubers. Plant Mol Biol 16:1009–1018
17. Ikeda T, Abe K, Ota A, Ikenoue T (1999) Heat shock protein 70 and heat shock cognate protein 70 messenger ribonucleic acid induction in the brains, hearts, and livers of neonatal rats after hypoxic stress. Am J Obstet Gynecol 180: 457–461
18. Nishizawa J, Nakai A, Komeda M, Ban T, Nagata K (2002) Increased preload directly induces the activation of heat shock transcription factor 1 in the left ventricular overloaded heart. Cardiovasc Res 55:341–348
19. Smith S, Stuber F, Schroeder S (2001) Protective functions of intracellular heat-shock protein (HSP) 70-expression in patients with severe sepsis. Intensive Care Med 27:1835–1841
20. Chang J, Wasser JS, Cornelussen RN, Knowlton AA (2001) Activation of heat-shock factor by stretch-activated channels in rat hearts. Circulation 104: 209–214
21. Ruddock LW, Klappa P (1999) Oxidative stress: protein folding with a novel redox switch. Curr Biol 9:R400–402
22. Arrigo AP (1998) Small stress proteins: chaperones that act as regulators of intracellular redox state and programmed cell death. Biol Chem 379:19–26
23. Snoeckx LH, Cornelussen RN, Van Nieuwenhoven A, Reneman RS, Van Der Vusse GJ (2001) Heat shock proteins and cardiovascular pathophysiology. Physiol Rev 81:1461–1497
24. Wong HR (1998) Potential protective role of the heat shock response in sepsis. New Horiz 6:194–200
25. Rutherford SL, Lindquist S (1998) Hsp 90 as a capacitor for morphological evolution. Nature 396:336–342
26. Rakitsky VN, Koblyakov VA, Turusov VS (2000) Nongenotoxic (epigenetic) carcinogens: pesticides as an example. A critical review. Teratog Carcinog Mutagen 20:229–240
27. Weber LP, Janz DM (2001) Effect of beta-naphthoflavone and dimethylbenz[a]anthracene on apoptosis and HSP70 expression in juvenile channel catfish ovary. Aquat Toxicol 54:39–50
28. Salminen WF Jr, Voellmy R, Roberts SM (1996) Induction of Hsp 70 in HepG2 cells in response to hepatotoxicants. Toxicol Appl Pharmacol 141:117–123
29. Mairesse N, Delhaye M, Galand P (1990) Enhanced expression of a 27 kD protein during diethylnitrosamine-induced hepatocarcinogenesis in rats. Biochem Biophys Res Commun 170:908–914
30. Lindeman B, Skarpen E, Huitfeldt HS (1998) Stress protein expression in rat liver during tumour promotion: induction of heat-shock protein 27 in hepatocytes exposed to 2-acetylaminofluorene. Carcinogenesis 19:1559–1563
31. Calabrese V, Renis M, Calderone A, Russo A, Reale S, Barcellona ML, Rizza V (1998) Stress proteins and SH-groups in oxidant-induced cellular injury after chronic ethanol administration in rat. Free Radic Biol Med 24:1159–1167

32. Warburg O (1956) "On The Origin of Cancer Cells." Science 123:309–314
33. Weber MJ, Evans PK, Johnson MA, McNair TF, Nakamura KD, Salter DW (1984) Transport of potassium, amino acids, and glucose in cells transformed by Rous sarcoma virus. Fed Proc 43:107–112
34. Marchand MJ, Maisin L, Hue L, Rousseau GG (1992) Activation of 6–phospho-fructo-2-kinase by pp60v-src is an indirect effect. Biochem J 285:413–417
35. Rady P, Arany I, Bojan F, Kertai P (1980) Effect of carcinogenic and non-carcinogenic chemicals on the activities of four glycolytic enzymes in mouse lung. Chem Biol Interact 31:209–213
36. Zhou R, Vander Heiden MG, Rudin CM (2002) Genotoxic exposure is associated with alterations in glucose uptake and metabolism. Cancer Res 62: 3515–3520
37. Beck EG (1976) Interaction between fibrous dust and cells in vitro. Ann Anat Pathol (Paris) 21:227–236
38. Bhat HK (2002) Depletion of mitochondrial DNA and enzyme in estrogen-induced hamster kidney tumors: a rodent model of hormonal carcinogenesis. J Biochem Mol Toxicol 16:1–9
39. Schmidt JV, Bradfield CA (1996) AH receptor signaling pathways. Annu Rev-Cell Dev Biol 12:55–89

Epidemiology: Aging as the Main Cause of Cancer

CHAPTER 10

With the notable exception of mortality among children and young adults the overall cancer death rate has been almost constant in the past 30 years, both in Europe and in the United States. In the meantime there has been a marked change in the site of primary cancer (increase in lung cancer, brain tumor, and melanoma), while there has been an unexplained decrease in cancer of the stomach, head and neck, and esophagus.
The aim of this chapter is to reinforce what has been known for years, namely, that the primary risk factor for cancer is aging and that there are complex interactions and competition between different diseases.
The aged organism is "a system near critical state". As the whole body ages, most carcinogens may simply have an incremental effect. Targeting of different organs causes a shift of primary cancer sites.

Modern epidemiology has made possible the identification of multiple risk or preventive factors for cancer. But, with the notable exception of stopping smoking, most interventions have failed to curb overall cancer mortality. There are two main reasons. First, most clinical trials have focused on a specific tumor type and may lack the ability to show a decrease in overall mortality. Second, and perhaps more important, large studies have shown a decrease in some types of primary cancer and, to the surprise of their authors, an increase in other tumor subtypes (1, 2).

Aging as the Main Reason for Cancer

The purpose of this chapter is to reinforce what has been known for years, namely, that the primary risk factor for cancer is aging and that there are complex interactions and competition between different diseases. Epidemiology deals mostly with age-standardized mortality rates, thereby somehow taking into account aging but also blurring its impact. A striking link exists between advanced age and increased incidence of cancer (3, 4). The clinical incidence of different cancers is spread through the human life span, but cancer is nevertheless mostly a disease of the elderly; two thirds of all carcinomas are diagnosed after the age of 70 years.

It appears that the overall cancer death rate has been almost constant in the past 30 years, both in Europe and in the United States, with the notable exception of mortality among children and young adults (5). In the meantime, there has been a marked shift in the distribution of tumor sites with an obvious rise in mortality by lung cancer, a less striking rise of melanoma or brain tumor, and a concomitant decrease in stomach, esophageal, and head and neck carcinoma. Smoking explains the increased lung cancer death rate, but the reasons for the other variations remain largely unknown. Furthermore, it appears that the overall cancer mortality rate varies by less than 20% in most developed countries (Table 10.1).

The fact that cancer incidence at specific sites is linked to nutrition is well known (6). Some cancer sites are associated with overnutrition (e.g., those of the colon, breast, and prostate) (6). The question still unanswered is whether an optimal lifestyle would increase life

Table 10.1. Overall cancer mortality in different developed countries for men and women (data from http://www-depdb.iarc.fr/who/menu.htm)

Country	ASR[a] (men)	ASR (women)
United States of America	148.38	104.38
China: urban and rural areas	152.53	84.33
Austria	154.65	92.86
France	185.53	84.49
Germany	162.61	99.79
Netherlands	173.57	106.22
Portugal	159.45	82.88
Spain	170.62	75.71
United Kingdom	157.21	111.93
Australia	146.67	93.15

[a] An age-standardized rate (ASR) is a summary measure of a rate that a population would have if it had a standard age structure. Standardization is necessary when comparing several populations that differ with respect to age, because age has such a powerful influence on the risk of cancer.

expectancy and decrease the overall cancer death rate. The answer is not straightforward, in particular at the most advanced age.

Studying the migration of populations suggests that change of diet decreases the incidence of some tumors but increases others. An example is the study of migrants from the former Soviet Union who arrived in Israel (7). For cancers of the digestive system, there is evidence of a marked reduction in the risk of stomach cancer, but in the same time period there is an increase in colon cancer. European immigrants to Sao Paulo were compared with those in the Brazil-born population and with those in their countries of origin. They had increased rates of oropharyngeal, esophageal, cervical, and breast cancer and decreased rates of lung cancer (8). Similarly, Japanese men in Hawaii show a decrease in stomach cancer rates but an increase in colon cancer rates (9).

In fact, diet and lifestyle are likely to have a definite but limited effect on overall cancer mortality. One example is that of Seventh-Day Adventists in California (10). By religious belief, Adventists do not smoke and do not consume alcohol or pork, with approximately one-half adhering to a lactovegetarian lifestyle. Standardized morbidity ratios (SMRs) for all cancer sites were calculated and cancer incidence rates in this population were compared with an external reference population. For all cancer sites combined in men, the SMR was lower in the Adventists (SMR=0.73). The SMR was also lower in men for most individual cancer sites. However, the prostate cancer risk was higher. Similar data were obtained by studying Norwegian Seventh-Day Adventists (11). The SMR in this group was 78.

This dependence of carcinogenesis on age is not limited to humans. One of the most constant pathological observations in senescent animals of all species is an increase in tumor frequency (12). In experiments with various carcinogenic agents (chemical, radiation, hormonal), it has been established that the sensitivity of different tissues to carcinogens changes with aging (13). Low doses of *N*-methyl-*N*'nitrosourea induce pancreatic carcinoma in aged mice but not in younger animals (14). Similarly, a low dose of DMBA is carcinogenic only in older rats (15). The predominant hypothesis is that old cells are more vulnerable to pathology and disease than young cells. But this may not be the reason for increased incidence of cancer in the elderly.

The Aged Organism as a System Near Critical State

Most mathematical models describing the evolution of mortality figures use the concept of death probability (16). When summarizing death statistics through this unique parameter, one implicitly makes the assumption that death events are independent from one individual to another. This hypothesis has profound consequences as it implies a "Gaussian" behavior of fluctuations in death statistics, which may not be correct. In order to verify the validity of this assumption, Filoche (16) studied French cancer death statistics between 1978 and 1996. The fluctuations, for every age bracket, were computed and then compared to the expected Gaussian fluctuations that should emerge from a simple model of death probability. The fluctuations were in close agreement with a Gaussian model up to age 35–40 years (Fig. 10.1). After 40 years of age, the fluctuations are much greater and cannot be explained by a model where every individual would have a given "probability of death."

This unexpected result suggests that cancer death statistics have the same mathematical signature as those of very different domains,

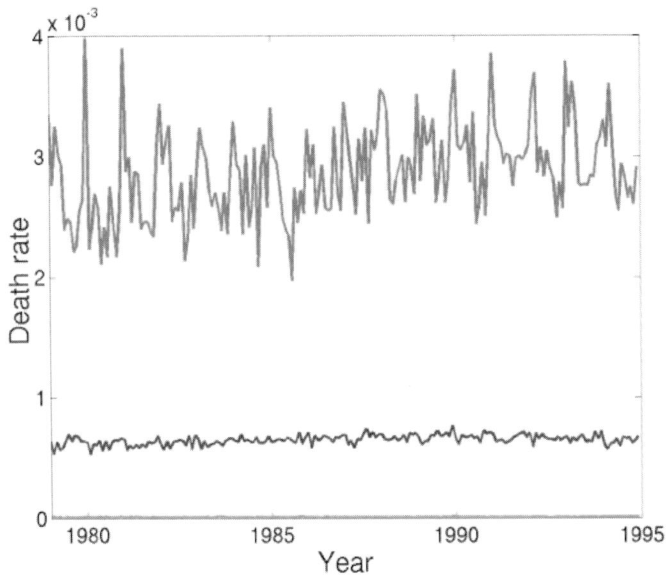

Fig. 10.1. Increased monthly fluctuation in mortality among older men. Monthly death rate by cancer for men's deaths between 1978 and 1995 for the age brackets 30–35 (*green*), 60–65 (*blue*), and 90–95 years (*red*). There is a strong increase of relative fluctuation with age, even though the monthly number of deaths increases (see [16])

such as seismology (distribution of earthquakes), fracture in material sciences, or "physics of the sand pile." The statistical fluctuations in the study of physical systems near critical states are much greater and more sensitive to external perturbations than in classical systems in physics, which are at or near equilibrium. Cancer acts on an aged organism that can be considered as a "system near a critical state." Thus, the global interaction between the disease and the organism can no longer be summarized in a single scalar quantity like the death probability, and the great fluctuations of death events observed in the statistics are a signature of this criticality.

This approach is confirmed by factorial analysis. This mathematical approach has been developed mostly for the field of finance and more recently for the medical field (17, 18). It compares the changes of incidence during a period of time and the evolution of one disease compared to another. A factorial analysis of cancer death and site-specific cancer death between 1970 and 1995 was performed. It shows little change in overall cancer mortality, with much greater differences in evolution of site-specific mortality (18).

The obvious question is why this dramatic contrast between stable overall cancer mortality and the great changes in the site-specific mortality. It is possible that as the whole body ages, most carcinogens simply have an incremental effect. In fact, carcinogens have, at the tissue and the cellular level, effects similar to those of normal aging.

References

1. Vainio H (1999) Chemoprevention of cancer: a controversial and instructive story. Br Med Bull 55:593–599
2. No authors listed (1994) The effect of vitamin E and beta carotene on the incidence of lung cancer and other cancers in male smokers. The Alpha-Tocopherol, Beta Carotene Cancer Prevention Study Group. N Engl J Med 330: 1029–1035
3. DePinho RA (2000) The age of cancer. Nature 408:6809
4. Macieira-Coelho A (2001) Neoplastic disease through the human life span. Biogerontology 2:179–192

5. Bailar JC, Gornik HL (1997) Cancer Undefeated. N Engl J Med 336:1569–1574.
6. Hill MJ (1997) Nutrition and human cancer. Ann N Y Acad Sci 833:68–78
7. Iscovich J, Howe GR (1998) Cancer incidence patterns (1972–91) among migrants from the Soviet Union to Israel. Cancer Causes Control 9:29–36
8. Bouchardy C, Khlat M, Mirra AP, Parkin DM (1993) Cancer risks among European migrants in Sao Paulo, Brazil. Eur J Cancer 29A:1418–1423
9. Stemmermann GN, Nomura AM, Chyou PH, Kato I, Kuroishi T (1991) Cancer incidence in Hawaiian Japanese: migrants from Okinawa compared with those from other prefectures. Jpn J Cancer Res 82:1366–1370
10. Mills PK, Beeson WL, Phillips RL, Fraser GE (1994) Cancer incidence among California Seventh-Day Adventists, 1976–1982. Am J Clin Nutr 59:1136S–1142S
11. Fonnebo V (1992) Mortality in Norwegian Seventh-Day Adventists 1962–1986. J Clin Epidemiol 45:157–167
12. Zimmerman JA, Carter TH (1989) Altered cellular responses to chemical carcinogens in aged animals. J of Gerontology 44:19–24
13. Anisimov VN (1985) Cancer and aging. Arch Geschwulstforsch 55:421–428
14. Zimmerman JA, Trombetta LD, Carter TH, Weisbroth SH (1982) Pancreatic carcinoma induced by N-Methyl-N'Nitrosourea in aged mice Gerontology 28:114–120
15. Russo J, Russo IH (1996) Experimentally induced mammary tumors in rats. Breast Cancer Res Treat 39:7–20
16. Filoche M, Schwartz L. Anomalous fluctuation in cancer statistics. Med Hypothesis (in print)
17. Jambu M (1989) Exploration statistique et informatique des donnnées. Dunod, Paris, France
18. Summa M,Goupil F, Cazes P, Sasco A, Steayert JM, Vautrain F, Schwartz L. Analyse factorielle de la mortalité par cancer. Revue de Statistique Appliquée (in print)

Aging Causes Inflammation and Fibrosis Mimicking the Impact of Carcinogens

CHAPTER 11

From the early 1900s until the mid 1960s, it was believed that aging had little to do with intracellular events and research on aging focused on extracellular phenomena. It was believed that normal cells given optimum environmental conditions had an unlimited capacity to replicate and function. Progress in the area of cell culture called into question, for a while, the essential role of these extracellular alterations. In fact, aging is directly responsible for inflammation and changes in tissue architecture mimicking the impact of carcinogens.

The Koch bacillus is responsible for tuberculosis. There is no tuberculosis without it. When we eliminate the Koch bacillus, tuberculosis disappears. The same logic does not apply to cancer. Carcinogens are not the sole origin of cancer; they only seem responsible for its premature appearance. A few rare lung cancer patients have never smoked; some patients suffering from mesothelioma have not been exposed to any detectable concentration of asbestos.

Asbestos seems to be responsible for 80% of cases of mesothelioma. The other 20% of cases are older patients for whom not even the most rigorous questioning (we call it "police interrogation" in our jargon) finds the least exposure to asbestos (1, 2). Once inhaled, asbestos fibers cause pleural plaques. This is in fact fibrosis localized in the pleura, secondary to chronic inflammation produced by these sharp fibers. Pleural plaques remain asymptomatic for a long time. Sometimes, however, they become cancerous and squeeze the lungs to the point of suffocating the patient. Pleural plaques are rare in young people; when they exist, questioning reveals exposure to asbestos. In aged patients on the other hand, pleural plaques are frequent and deaths from other causes prevent them from developing into aggressive cancers. Meticulous autopsy reveals such plaques in almost 50% of elderly patients who have not been exposed to asbestos (3).

Just as asbestos is responsible for most mesotheliomas, smoking causes lung cancer. The great majority of young patients with lung cancer are either smokers or ex-smokers. This link is more tenuous in the aged (4).

Aging Causes Inflammation

Aging is associated with increased levels of chronic inflammation and with increased inflammatory activity reflected by increased circulating levels of TNF-alpha, IL-6, cytokine antagonists, and acute phase proteins in vivo (5). These changes in the extracellular matrix disturb intercellular communication.

Just as in inflammation, the most profound losses in aging are those that involve coordination between cells. Muscle strength decreases by 25% between the age of 30 and 80 years, while the coordination between nerve and muscle decreases by 50% (6).

Epidemiologic studies suggest that chronic low-grade inflammation in aging promotes atherosclerosis and causes age-associated disorders such as Alzheimer's disease, macular degeneration, and type 2 diabetes leading to an increased mortality risk. This inflammation caused by aging plays a major role in the decline in immune function and lean body mass (5, 7).

Aging: Changes in Tissue Architecture – the Example of the Skin

Architectural changes caused by aging can be seen every morning when looking in the mirror. Who has not noticed the difference between the soft skin of a child and the much rougher skin of the elderly? Atrophy, wrinkling, sagging, and laxity are the most obvious sign of old skin (8). A good doctor can estimate a patient's age within 1 or 2 years simply by looking at the person. This estimate does not involve sophisticated biological tests but is based simply on noticing the extent of the wrinkles.

These architectural changes are caused by the alteration of the connective tissue underlying the epithelium, leading to folds and wrinkles (8). Wrinkles are the sum of age-related changes occurring in the mesenchymal cells and the supporting macromolecular structures. Aging changes the architecture of the fibroblast and the surrounding stroma. The nucleus increases in size and becomes rounder with age (9); there is a change in the organization of the cytoplasmic filaments and of the extracellular matrix (10). The fibroblasts extracted from older volunteers show a disordered actin cytoskeleton, a reduced ability to contract collagen gels, and limited ability to migrate (9). In vitro and in vivo experiments show that cellular and extracellular aging is part of the same process (8).

There are many changes that do not compromise health. No one has ever died of wrinkled skin or gray hair. But the same changes occur in every organ, resulting in multiple architectural changes (11). Gastronomy provides a clear example of change in the extracellular matrix. The flesh of a young animal is more tender than that of an old animal, no matter what the cut of meat.

An aging liver presents architectural anomalies, with reduced sinusoidal fenestration, collagen deposition, and neoangiogenesis (12). Arteries also become more rigid with age. When the pulse is taken, the cardiac ejection time shock wave is more intense. The reason for this is, once again, extracellular (13, 14).

Microgravity: A Model of Aging

Mechanical forces play an important role in the architectural deterioration of the dermis (8, 9). The question is whether these physical changes are sufficient to explain aging.

One example comes to us from the sky. We have all seen pictures of young cosmonauts back on earth, sitting in their chairs, unable to stand up. Weightlessness (microgravity, as astronauts call it) has brought about a disease of old age: incapacitating osteoporosis with muscular atrophy and bone loss (15).

Studies from the Skylab missions have demonstrated that during pace flight animal and human cells appear to undergo changes similar to aging. Basic cellular functions such as electrolyte concentration, cell growth rate, glucose utilization, bone formation, response to growth stimulation, and exocytosis are modified in microgravity. Many of the physiological changes seen in humans in spaceflight may originate from dysfunction of basic biological mechanisms caused by microgravity. Aging humans share many of the symptoms seen in astronauts during spaceflight. These include reduced cardiac function, loss of bone, reduced immune response, and orthostatic hypotension (15).

Aging and Carcinogenesis

The leathery skin of a sailor ages faster than the soft face of an urban secretary. It is on the sailor's skin, not on that of the secretary, that basocellular and spinocellular carcinomas caused by the sun appear early in life. Carcinogens, in this case ultraviolet rays, are responsible for both early aging of the skin and the appearance of skin cancers. The secretary is not exempt from the possibility of skin cancer but, in general, it will appear later.

Ultraviolet irradiation speeds up skin aging and induces skin carcinogenesis. Photoaged skin is characterized by early wrinkles as well as loss of skin tone and resilience (16). Like skin that has aged prematurely, photoaged skin displays alterations in extracellular matrix with accumulation, disorganized elastin and fibrillin, and a severe loss of interstitial collagens, the major structural proteins of the dermal connective tissue. These alterations are not specific and are seen, albeit in a milder form, in older people not exposed to high doses of ultraviolet irradiation.

Factors as diverse as ultraviolet radiation, wounds, infections, traumatisms, anoxia, cigarette smoke, physical stress, and hormonal status play a role and have been termed "factors of aging." All these factors share as a common feature: the ability to directly or indirectly change the extracellular environment of the cell and increase cancer incidence (16, 17).

References

1. McDonald JC, Armstrong BG, Edwards CW, Gibbs AR, Lloyd HM, Pooley FD, Ross, DJ, Rudd RM (2001) Case-referent survey of young adults with mesothelioma: I. Lung fibre analyses. Ann Occup Hyg 45:513–518
2. Neumann V, Gunthe S, Mulle KM, Fischer M (2001) Malignant mesothelioma–German mesothelioma register 1987–1999. Int Arch Occup Environ Health 74:383–395
3. Francis D, Jussuf A, Mortensen T, Sikjaer B, Viskum K (1977) Hyaline pleural plaques and asbestos bodies in198 randomized autopsies. Scand J Respir Dis 58:193–196
4. Franceschi S, Bidoli E (1999) The epidemiology of lung cancer. Ann Oncol 10:S3–6
5. Grimble RF (2003) Inflammatory response in the elderly. Curr Opin Clin Nutr Met Care 1:21–29
6. Shock N (1983) Aging of regulatory mechanism. In: Cape RDT, Coe RM, Rossman I (eds) Fundamentals of geriatric Medecine. Raven Press, New York, pp 51–62
7. Bruunsgaard H, Pedersen M, Pedersen BK (2001) Aging and proinflammatory cytokines. Curr Opin Hematol 8:131–136
8. Lapiere CM (1990) The aging dermis: the main cause for the appearance of 'old' skin. Br J Dermatol 122:5–11
9. Pienta KJ, Getzenberg RH, Coffey DS (1992) Characterization of nuclear morphology and nuclear matrices in aging human fibroblasts. Mech Aging Dev 62:13–24

10. Van Gansen P, Van Lerberghe N (1988) Potential and limitations of cultivated fibroblasts in the study of senescence in animals. A review on the murine skin fibroblasts system. Arch Gerontol Geriatr 7:31–74
11. Reed MJ, Ferara NS, Vernon RB (2001) Impaired migration, integrin function, and actin cytoskeletal organization in dermal fibroblasts from a subset of aged human donors. Mech Aging Dev 122:1203–1220
12. Jansen PL (2002) Liver disease in the elderly. Best Pract Res Clin Gastroenterol 16:149–58
13. Silver FH, Horvath I, Foran DJ (2001) Viscoelasticity of the vessel wall: the role of collagen and elastic fibers. Crit Rev Biomed Eng 29:279–301
14. O'Rourke MF (1999) Mechanical principles. Arterial stiffness and wave reflection. Pathol Biol (Paris) 47:623–633
15. Hughes-Fulford M (1991) Altered cell function in micro gravity. ExpGerontol 26:247–256
16. Wlaschek M, Tantcheva-Poor I, Naderi L, Ma W, Schneider LA, Razi-Wolf Z, Schuller J, Scharffetter-Kochanek K (2001) Solar UV irradiation and dermal photoaging. J Photochem Photobiol B 63:41–51
17. Giacomoni PU, Rein G (2001) Factors of skin aging share common mechanisms. Biogerontology 2:219–229

Inhibition of Metabolism Slows Aging and Decreases Cancer Incidence

CHAPTER 12

In organisms ranging from yeast to mice, mutations in glucose transport pathways decrease the glucose uptake and extend life span. Similarly, restriction of the number of calories extends longevity. In rodents and primates, caloric restriction decreases the levels of plasma glucose, attenuates inflammation, and postpones both senescence and cancer without irreversible side effects.

The most striking difference in cancer incidence is between sexes and social classes. Cancer incidence is higher among manual workers than upper middle class intellectuals. This difference cannot simply be explained by environmental factors like smoking or excessive drinking. Manual workers experience more rapid aging and a concomitant increased incidence of cardiovascular or neurodegenerative diseases. Working hard means having a higher metabolism. This increased oxygen consumption may explain both aging and increased cancer incidence.

Lower Metabolism May Explain Slower Aging and Decreased Cancer Incidence in Women

In all developed countries, life expectancy at birth is higher for women than for men (1). In Sweden, the country which first implemented reliable statistics, female life expectancy has exceeded that of men since 1751 (1, 2). Average female life expectancy now exceeds 80 years in most developed countries (3). Women outlive men by 5–9 years (3). The principal reason for this widening gap has been the reduction in mortality of women at higher ages (Fig. 12.1). This difference is not limited to cancer but appears valid for most causes of death. The gender difference is constant in cardiovascular disease, cancer, and dementia. Although there are exceptions to the rule, the difference in mortality between men and women (gender gap) is a general phenomenon observed in the animal kingdom (2, 4).

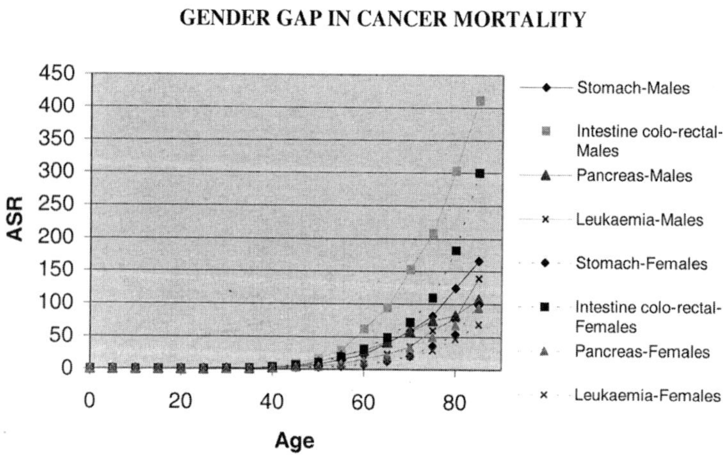

Fig. 12.1. Cancer death rate is strongly linked to age. There is a gender gap even for primary sites (pancreas, blood, colon, and stomach) not obviously related to smoking and drinking

Every year, in every country, and for every tumor site (for those organs present in both sexes), the age-adjusted cancer incidence and death rate are lower for women, with the exception of thyroid cancer. Although several hypotheses have been formulated, there is still no clear answer to the gender gap. Biological, behavioral, and environmental factors are suggested as major contributors to the difference in cancer incidence between men and women (1, 2, 5). For example, to date, women have smoked less and drunk less alcohol than men. However, this gender gap is found in tumors caused by the environment, such as lung cancer, but also for tumors whose risk factors are not as well known (colon cancer) or even unknown (brain tumors) (Fig. 12.1).

The difference in incidence and mortality between men and women may simply reflect a difference in inbred metabolism. It is striking that this difference is not limited to cancer or even to adulthood. During the first year of human life, when behavior presumably does not obviously differ according to sex, there is a higher fetal mortality rate in baby boys, making an environmental explanation unlikely. In fact, the explanation for the gender difference is more likely to be decreased metabolism and therefore slower aging and lesser tissue damage for women than for men. It has been long known that newborn males are heavier than females. This differential of growth is established at a very early stage prior to the production of fetal hormone (6, 7). A study by Fog Petersen (6, 7) measuring fetal crown–rump lengths and biparietal diameter showed that female fetuses at 8–12 menstrual weeks were 1 day behind male fetuses and the difference rose to 6–7 days at term. There is increasing evidence that the differential growth starts within days after fecundation (8), probably because paternal Y chromosomal genes may be responsible for early cleavage divisions of the male embryo (9) and the resulting quicker growth rate. Even within the same sex, male or female, early growth rate has been suggested to be important for future cancer risk (10, 11). Increased cancer incidence among the young is associated with higher birth weights (11). Similarly, an increased risk of later renal cell cancer was observed among men with a birth weight of 3,500 g or more compared with men with a lower birth weight. There is an increased risk of ovarian cancer in girls exhibiting a tall stature in early childhood and high birth weight (12). Testicular cancer incidence peaks during early adulthood. Few risk factors have been established for testicular cancer, but evidence suggests that causal factors operate early in life, perhaps even in utero (13). There is an increased risk of testicular cancer among high (≥4,000 g) birth weight babies. Birth weight is also significantly associated with breast cancer risk (14).

Caloric Restriction Reduces and Delays Aging and Consequent Carcinogenesis

It has been known for 60 years that dietary caloric restriction, a 30% reduced diet for most animal life, is the only intervention conclusively and reproducibly shown to slow aging and maintain health and vitality in rodents or primates.

A caloric restriction diet (CR) lowers the metabolism. Height, weight, and body temperature are lowered, plasma insulin levels are reduced, and cholesterol and triglycerides are decreased. Blood pressure is below normal.

The inflammatory response of dieting animals is reduced (15). The proinflammatory cytokines NF-kappa B, IL-1 beta, IL-6, TNF-alpha, cyclooxygenase-2, and inducible NO synthase are attenuated by CR (16). In dieting animals, decreased collagen production and fewer architectural changes are observed (17). These animals age more slowly than their fully fed counterpart (18). Accordingly, there is a marked decrease in the incidence of a wide range of age-related diseases such as diabetes, cardiovascular problems, and spontaneous cancer.

Caloric restriction in rodents or primates also results in the reduction of chemically induced neoplasms (18).

Similar data are obtained when studying patients with anorexia nervosa (19). The overall cancer incidence among women with anorexia nervosa was reduced by 20% below that of the general population. The finding of this reduction in cancer risk among women with anorexia nervosa supports the theory that a low-energy diet may decrease carcinogenesis in humans (19).

Conclusion

Our hypothesis is that carcinogens induce glycolysis and thus carcinogenesis. If this were correct, mutations would not be the cause but the consequence of carcinogenesis. The limited data in the literature suggest this may be the case. Glucose depletion, as in caloric restriction, delays carcinogenesis (20).

Deficiency in the HIF gene results in decreased uptake of glucose. The subcutaneous injection of tumor cells deficient in HIF is then less likely to form tumor (21).

References

1. Giampaoli S (2000) Epidemiology of major age-related diseases in women compared to men. Aging Clin Exp Res 12:93–105
2. Trussel J (1995) Women's longevity. Science 270:720–721
3. Kinsella K (1992) Changes in life expectancy 1900–1990. Am J Clin Nutr 55:1196S-1202S
4. Leveille SG, Resnick HE, Balfour J (2000) Gender differences in disability: Evidence and underlying reasons. Aging Clin Exp Res 12:106–112
5. Tarin JJ (1997) Do the fastest concepti have a shorter life span? Human Reproduction 12:885–889
6. Clarke CA, Mittwoch U (1995) Changes in the male to female ratio at different stages of life. Br J Obstect Gynec 102:677–679
7. Mittwoch U (1993) Blastocysts prepare for the race to be male. Human reproduction 8:1550–1555
8. Pergament E, Fiddler M, Cho N, Johnson D, Holmgren J (1994) Sexual differentiation and preimplementation cell growth. Human reproduction 9:1730–1732
9. Ao A, Erickson RP, Winston RM, Handyside AH (1994) Transcription of paternal Y-linked genes in the human zygote as early as the pronucleated stage. Zygote 2:281–287

10. Bergstrom A, Lindblad P, Wolk A (2001) Birth weight and risk of renal cell cancer. Kidney Int 59:1110–1113
11. Ferguson P, Mohr L, Hoel D, Lipsitz S, Lackland D (2000) Possible relationship between birth weight and cancer incidence among young adults. Ann Epidemiol 10:471
12. Barker DJ, Winter PD, Osmond C, Phillips DI, Sultan HY (1995) Weight gain in infancy and cancer of the ovary. Lancet 345:1087–1088
13. Akre O, Ekbom A, Hsieh CC, Trichopoulos D, Adami HO (1996) Testicular nonseminoma and seminoma in relation to perinatal characteristics. J Natl Cancer Inst 88:883–889
14. Michels KB, Trichopoulos D, Robins JM, Rosner BA, Manson JE, Hunter DJ, Colditz GA, Hankinson SE, Speizer FE, Willett WC (1996) Birthweight as a risk factor for breast cancer. Lancet 348:1542–1546
15. Frame LT, Hart RW, Leakey JE (1998) Caloric restriction as a mechanism mediating resistance to environmental disease. Environ Health Perspect 106:313–324
16. Chung HY, Kim HJ, Kim KW, Choi JS, Yu BP (2002) Molecular inflammation hypothesis of aging based on the anti-aging mechanism of calorie restriction. Microsc Res Tech 59:264–272
17. Reiser K, McGee C, Rucker R, McDonald R (1995) Effects of aging and caloric restriction on extracellular matrix biosynthesis in a model of injury repair in rats. J Gerontol A Biol Sci Med Sci 50:B40–47
18. Roth GS, Ingram DK, Lane MA (2001) Caloric restriction in primates and relevance to humans. Ann N Y Acad Sci 928:305–315
19. Mellemkjaer L, Emborg C, Gridley G, Munk-Jorgensen P, Johansen C, Tjonneland A, Kjaer SK, Olsen JH (2001) Anorexia nervosa and cancer risk. Cancer Causes Control 12:173–177
20. Roth GS, Ingram DK, Lane MA (2001) Caloric restriction in primates and relevance to humans. Ann N Y Acad Sci 928:305–315
21. Williams K, Tefla B, Anley R, Sheridan M, Van der Kogel A, Harris AL, Stratford IJ (2002) A protective role for HIF-1 in response to redox manipulation and glucose deprivation: implications for tumorigenesis. Oncogene 21:282–290

CHAPTER 13

A Complication of Hypoxia: The Loss of Cell Polarity Explains Cancer Cell Proliferation

One of the hallmarks of cancer is its unrestrained growth. The multiplication of cancer cells is dependent on physical forces. Cancer invades preferentially soft tissues like glands or muscle rather than fascia or bone.
Cancer and normal cells may not react differently to mechanical forces, but as cancer cells can divide on a different plane, they are not exposed to the same physical constraints. It is likely that new physical constraints due to tissue disruption coupled with a loss of intercellular junctions (a consequence of anaerobic glycolysis) are enough to explain loss of contact inhibition and unrestrained cell proliferation.

Normal epithelial cells are organized along a structural axis (1) which allows cell adhesion to the mesenchyme on one side and epithelium function on the lumen side. This functional polarity is maintained by the cytoskeleton, by vesicular trafficking that proceeds along the microtubules (2), by organelles such as the centrosomes and, in general, by the interpretation of cues coming from the surrounding tissue. One side of the epithelial cell lies on the basement membrane, one side faces the lumen, and there are cells on the other surrounding sides. This normal epithelial cell can only divide on one plane. The cell cannot jump toward the lumen and create a new cell layer. Cell multiplication will be responsible for the epithelial elongation of its tesselation. The uniaxial stress generated by cell division may only fold or elongate the tissue.

This is confirmed in vitro. Normal cells in a Petri dish multiply until they come into contact with the edges of the dish. Upon contact, growth stops: therefore, there is only one layer of cells.

Loss of Cell Polarity: A Consequence of Hypoxia

This functional cell polarity is maintained both by gap junctions and by the microtubules that originate in the centrosome and play a key role in the architecture of the epithelium (3). Gap junctions, the cytoskeleton, and the intercellular junctions are necessary for mechanotransduction across the cell surface and at multiple locations inside the cell (3, 4).

Hypoxia causes tissue disorganization and decreased cell-to-cell attachment. Hypoxia inhibits the gap junction and decreases cell-to-cell interaction (5–8). This facilitates blood cells reaching the infection site. By-products of anaerobic glycolysis such as proinflammatory cytokines TNF-alpha or IL-1 inhibit gap junctions (9).

Inhibition of gap junction intercellular communication occurs both in inflammation and during carcinogenesis (10). Experimental data correlate the loss of gap junction and malignant progression (10–12). Nongenotoxic carcinogens frequently inhibit gap junction intercellullar communication or tubulin polymerization resulting in the loss of cell polarity (13).

Because of loss of polarity, tumor cells have lost this contact inhibition. They can jump and create a new cell layer. They can multiply on another plane and spread over several layers. Anatomopathologic examinations reveal the loss of cell polarity in cancers. An epithelial cell faces a certain direction. For instance, in bronchial epithelial cells, the ciliary zone faces the bronchial light. Like an escalator, this zone allows sputum, as well as bacteria, to come up from the alveoli and be expelled. When cancer is present, the cell no longer has this privileged orientation, mucus vacuoles are no longer anchored to the basal membrane but are scattered everywhere, and the cilia have disappeared. The mitoses no longer have a preferred axis.

Galen's Crab Yesterday, Fractal Growth Today

At the macroscopic level, cancer is always characterized by specific growth patterns such as expansive (cauliflower-like), infiltrative, or radiating patterns (14). It is the irregular dendritic-like pattern of the melanoma that the dermatologist looks for. In gynecology, the fractal deposits of calcium on mammographs are the sign of early breast cancer (14–18) (Fig. 13.1). Since the infiltrative-radiating pattern occurs in every malignant subtype, studies were essentially motivated by the design of new diagnostic tools (especially image analysis) (19).

Carcinomas arise in the epithelium lining the lumen of organs. Cancer is often preceded by a continuous sequence of events over time, which modifies the architecture of the epithelium. The evolution starts with metaplasia and progresses through dysplasia up to in situ carcinoma and then invasive carcinoma.

The normal epithelium is well ordered, with a single row of cells in perfect alignment. During carcinogenesis, there is progressive thickening of the tissue, accompanied by progressive disorder inside the increased number of cellular layers (metaplasia and dysplasia). The layers close to the mesenchyme still show a correlation between neighboring cells, while the top layers show no apparent order. The tendency for cells to be aligned with their neighbors decreases in the top layers.

In carcinoma in situ, the number of cell layers is further increased. Individual cells have lost their polarity. The cell division of cancerous cells is random. There is still no effraction of the basement membrane and invasion of the underlying mesenchyme.

Fig. 13.1a–c. These images are similar to the dendritic growth of cancer, but they correspond to the fractal deposits of copper on glass. The same equation can be used for modelized cancer growth and fractal deposits of electrolytes (Courtesy Vincent Fleury: PMC. Ecole Polytechnique Palaiseau)

In invasive carcinoma, disordered tissue grows toward the mesenchyme. In this invasive system, the shape of the cancer may be anything from a few buds or fingers (as physicists would probably call them) to a very complex branching structure, which is impossible to separate from normal tissue.

Galen's Crab Is a Simple Consequence of the Loss of Cell Polarity

These perceived links between loss of cellular polarity and architectural anomalies can be demonstrated by using the theoretical tools perfected by mathematicians and physicists.

While it has long been difficult to even describe cancer morphologies, there are now new concepts, such as fractal geometry (20), which can be used to describe the irregular structures of invasive tumors. Computer modelization of fractal growth has made it possible to study rough or irregular shapes (21, 22). The models seem very attractive in that they are easily able to generate very complex, rough, or invasive patterns, whose shapes have "universal" features.

The simplest model of growth was proposed 40 years ago by Eden (23). More recently, two important families of models have been introduced: first, the invasion percolation (IP) (24), and second, the diffusion limited aggregation (DLA) model (25). These three families of models (Eden, IP, DLA) and many related ones have generated an enormous body of work by physicists, but their very existence has been largely unknown to most biologists and cancer researchers (19).

Nevertheless, these models have been used in such fields as vasculogenesis (26–28), bacterial colony growth (29), coral growth (30), plant growth (31), tumor growth (19, 32), among others. In search of a correlation between shape and pathology, some studies have applied fractal analysis to the study of specific tumor shapes, but very few have tried to understand the tumor growth pattern itself in the context of fractal or irregular growth(14–19).

What these mathematical models tell us is that all we need to get such a complex fractal shape is cellular division in another plane. The cancer cell must escape physical constraint and be able to jump toward the lumen. This ability to "jump" is called loss of cell polarity. With this sole hypothesis, these mathematical models of fractal growth are able to reproduce qualitatively the sequence of observed abnormalities of metaplasia, dysplasia, and carcinoma in situ and invasive carcinoma. The models are based on loss of cell polarity only, in cancerous cells, coupled with otherwise normal growth rate and epithelial behavior. These models show that, as the probability of a wrong plane of cell division is increased, a transition from a normal, well-stratified epithelium to an invasive, fractal, dendritic pattern is observed. This transition shows a sequence of morphologies in the following order as a function of loss of polarity: first, apparently normal but already diseased tissue, then metaplastic tissue followed by dysplastic tissue, and eventually carcinoma, first in situ then invasive.

Therefore, one is led to argue that isolated precancerous or cancerous cells exhibit a loss of polarity, while the invasive cancer as a whole exhibits an irregular or star-shaped structure in which the normal morphology of the organ is destroyed or lost. This heuristic reasoning identifies on the one hand disturbance of a local rule, and on the other hand, disturbance in macroscopic shape. It is typical in physics that a change in local rules between neighbors will induce a morphology transition acting on the whole and on long-range correlations. In this case, the transition seems to stem from a loss of short-range instruction about the direction of mitosis, which induces global misbehavior when these mitoses add up. It is reminiscent of a roughening transition, as known in physics (33), in that the surface either has long-range correlation (perfect alignment of entities: the atoms for a crystal, the cells for an epithelium) in one state or a disordered bulk and surface in the other state.

This computer simulation supports the idea that loss of cell polarity (an individual property of cells) can explain the transition between well-organized tissue and invasive cancer. This simulation,

Table 13.1. Consequence of loss of cell polarity

Microscopic	Macroscopic
Cell multiplication	Metaplasia
Loss of cell differentiation	Dysplasia
	Carcinoma "in situ"
	Invasive carcinoma

whose basic ingredients correspond to known cellular behaviors, points, at least on a qualitative level, to the existence of a link between these well-known clinical features (Table 13.1).

1. Both metaplasia and dysplasia are premalignant changes.
2. The rate of spontaneous regression is lower for dysplasia than for metaplasia.
3. High-grade dysplasias (i.e., with the most severe loss of cellular polarity) are the most likely to evolve into a carcinoma in situ.
4. The greater the loss of polarity, the higher the tumor grade, and the more aggressive and invasive the cancer.
5. The loss of cellular polarity is specific to cancer growth in vitro and one of the definitions of malignant transformation.
6. In the vicinity of genuine tumors, smaller dormant clones may co-exist within the neighboring metaplasia (34).

Loss of Polarity Explains Unrestrained Cancer Growth

The invasiveness of neoplastic cells is dependent on physical forces. Cancer invades preferentially soft tissues like glands or muscle rather than fascia or bone. In vivo, chronic physical irritation can alter cancer cell growth.

The effects of mechanical forces on cancer cells have not been studied systematically in vitro. A few cancer cell lines have been investigated primarily to model normal processes (35), but not to investigate carcinogenesis. As in normal cells, mechanical deformation induces cell proliferation (35). Cell proliferation of a colon carcinoma cell line, HCT116, is increased by 30% after 2 days of deformation (30 cycles/min). However, solid stress (45–120 mmHg) inhibits multicellular spheroid tumor (36).

Cancer and normal cells do not appear to react differently to mechanical forces; however, because cancer cells can divide on a different plane, they elude constraints and creep between obstacles.

Unrestrained cell proliferation can be explained simply by the loss of cell-to-cell junction and change in physical constraints. This loss of cell polarity and change in constraints are both a consequence of inflammation.

References

1. Locke D (1998) Gap junction in normal and neoplastic mammary gland. J Pathol 186 : 343–349
2. Lingle W, Lutz W, Ingle J, Maihle N, Salisbury J (1998) Centrosome hypertrophy in human breast tumors. Proc Nat Acad Sci USA 95 : 2950–2955
3. Donahue HJ (2000) Gap junctions and biophysical regulation of bone cell differentiation. Bone 26 : 417–422
4. Wang N, Butler J, Ingber D (1993) Mechanotransduction across the cell surface and through the cytoskeleton. Science 260 : 1124–1128
5. Nishida M, Futami S, Morita I, Maekawa K, Murota SI (2000) Hypoxia-reoxygenation inhibits gap junctional communication in cultured human umbilical vein endothelial cells. Endothelium 7 : 279–286
6. Zhang YW, Morita I, Zhang L, Shao G, Yao XS, Murota S (2000) Screening of anti-hypoxia/reoxygenation agents by an in vitro method. Part 2: Inhibition of tyrosine kinase activation prevented hypoxia/reoxygenation-induced injury in endothelial gap junctional intercellular communication. Planta Med 66 : 119–123
7. Martinez AD, Saez JC (2000) Regulation of astrocyte gap junctions by hypoxia-reoxygenation. Brain Res Brain Res Rev 32 : 250–258
8. Peters NS (1995) Myocardial gap junction organization in ischemia and infarction. Microsc Res Tech 31 : 375–386
9. Hu J, Cotgreave IA (1997) Differential regulation of gap junctions by proinflammatory mediators in vitro. J Clin Invest 99 : 2312–2316

10. Rosenkranz HS, Pollack N, Cunningham AR (2000) Exploring the relationship between the inhibition of gap junctionnal intercellular communication and other biological phenomena. Carcinogenesis 21:1007–1011
11. Yamasaki H, Mesnil M, Omori Y, Mironov N, Krutovskikh V (1995) Intercellular communication and cancer. Mut Res 333:181–188
12. Locke D (1998) Gap junction in normal and neoplastic mammary gland. J Pathol 186:343–349
13. Combes RD (2000) The use of structure-activity relationships and markers of cell toxicity to detect non-genotoxic carcinogens. Toxicol InVitro 14:387–389
14. Sedivy R (1996) Fractal tumors: their real and virtual images. Wien Klin Wochenschr 108:547–551
15. Cross S (1997) Fractals in pathology. Journal of Pathology 182:1–8
16. Smolle J (1998) Fractal tumor stromal border in a nonequilibrium growth model. Analyt Quant Cytol Histol 20:7–13
17. Losa GA (1995) Fractals in pathology: are they really useful? Pathologica 87:310–317
18. Baish JW, Jain RK (2000) Fractals and cancer. Cancer Research 60:3683–3688
19. Fleury V, Schwartz L (2003) Numerical investigation of the effect of loss of polarity on cancer invasiveness and geometry. Fractals 11:1–15
20. Mandelbrot B (1977) The fractal geometry of nature. Freeman and Co, San Francisco
21. Vicsek T (1992) Fractal Growth Phenomena, World Scientific (Singapore), Second Edition
22. Meakin P (1988) Aggregation phenomena. In: Domb C, Leibowitz JL (eds) Phase transitions and critical phenomena. Academic Press
23. Eden NM (1961) Proceedings of the Fourth Berkeley Symposium on Mathematics, Statistics and Probability, edited by F. Neyman, Vol. IV (University of California Press, Berkeley)
24. Stauffer D, Aharony A (1994) Introduction to Percolation Theory. Taylor and Francis, London
25. Witten TA, Sander LM (1981) Diffusion limited aggregation as a critical phenomenon. Phys Rev Lett 47:1400
26. Family F, Masters BR, Platt DE (1989) Fractal pattern formation in human retinal vessels. Physica D 38:98–103
27. Fleury V, Schwartz L (2000) Fractals, Modelisation of 3D microvasculature by interlaced DLA. Fractals 8:255–259
28. Bard J (1990) Morphogenesis. Cambridge University Press, Cambridge, Mass, 212–214, paragraph 6.6.1
29. Ben Jacob E, Schochet O, Tenenbaum A, Cohen I, Czirok A, Vicsek T (1994) Generic modelling of cooperative growth patterns in bacterial colonies. Nature 68:46–49
30. Kaandorp JA, Lowe CP, Frenkel D, Sloot Peter MA (1996) Effect of Nutrient Diffusion and Flow on Coral Morphology. Phys Rev Lett 77:2328–2331

31. Fleury V (1999) A possible connection between dendritic growth in physics and plant morphogenesis. C R Acad Sc Paris, série biologie 322:725–734
32. Landini G, Rippin JW (1995) Tumour shape and local connected fractal dimension analysis in oral cancer and pre-cancer. Fractal reviews in natural and applied sciences, 171–178, Chapman and Hall
33. Hurle TJ (1994) Handbook of crystal growth, Vol. 1. Elsevier Science, North Holland (Amsterdam)
34. Park IW, Wistuba I, Maitra A, Milchgrub S, Virmani AK, Minna JD (1999) Multiple clonal abnormalities in the bronchial epithelium of patients with lung cancer. J Natl Cancer Inst 91:1863–1868
35. Johnson F, Zhou M, Collins B, Huang J (2000) Mechanical deformation induces proliferation of human colorectal carcinoma cells. Int J Oncol 16:617–622
36. Helmlinger G, Netti PA, Lichtenbeld HC, Melder RJ, Jain RK (1997) Solid stress inhibits the growth of multicellular tumor spheroïds. Nat. Biotechnol 8: 778–783

The Loss of Cell Differentiation Is a Consequence of Glycolysis

CHAPTER 14

Respiration provides energy for differentiated human cells like the ones of the muscles, kidney, or heart. These cells are well differentiated and do not divide. Glucose is degraded into water and gas. Lesser-differentiated cells like bone marrow stem cells use both aerobic and anaerobic glycolysis. The switch from respiration to anaerobic glycolysis appears responsible for both cell proliferation and loss of cell differentiation.

During carcinogenesis, architectural changes (the epithelium increases in depth) and cytological modifications (increased mitoses and cellular pleomorphisms) coexist. Thus, there is coexistence of anomalies of different scales: the micrometer for the cell, the millimeter for architecture.

Cancer cells have lost most of their original cell differentiation. Breast cancer does not secrete milk, salivary gland tumors do not synthesize saliva. The reason for this loss of cellular differentiation remains largely unknown. The current predominant hypothesis explains cellular differentiation and its loss in cancer cells as an essentially intracellular genetic process.

However, it is possible or even probable that that the loss of cell differentiation such as seen in cancer results from a change in cellular environment.

Cancer Cell Differentiation

Most of the research has focused on the differences between tumor subtypes. Cancers are classified based on their cellular differentiation. Some cancers secrete keratin, just as normal skin cells do; others secrete mucus as glandular cells do. Often, these keratin or mucus deposits are visible at microscopic examination. More rarely, the use of an antibody is needed to detect tumoral secretions. A tumor which secretes mucus is an adenocarcinoma; a tumor secreting keratin is an epidermoid cancer. In rare cases, an adenosquamous cancer will secrete both mucus and keratin. Epidermoid or glandular differentiation has little influence on treatment or prognosis.

Adenocarcinomas and squamous cell carcinomas arise in specific primary tumor sites. Most cancers arising from the head and neck, the esophagus, or the cervix are squamous cell carcinomas. Like normal skin, these tumors synthesize keratin. On the other hand, cancers arising from the breast, the gastrointestinal tract, or the lining of the uterus are adenocarcinomas.

Most esophageal cancers are squamous cell carcinomas. The tumor arises equally from the proximal, the middle, or the distal third of the esophagus. The patients are usually heavy smokers and drinkers. This contrasts with adenocarcinomas of the esophagus. The localization of adenocarcinoma is different from that of squamous carcinoma. It is confined to the distal third of the esophagus and usually arises from the gastro-esophageal junction (G-E junction). The vast majority of patients with adenocarcinoma of the G-E junction have a history of acidic gastric reflux caused by hiatal hernia. Alcohol and cigarettes are not major risk factors for adenocarcinoma.

Mucus production and secretion is a common mechanism used by mammals to protect the underlying mucosae against various injuries (pollutants, pathogens, acidity) (1). The normal gastric epithelium is covered with a continuous layer of secreted mucus and bicarbonate (2). This gastric mucus can slow down acid diffusion and enable the formation of a pH gradient across the mucobicarbonate layer to protect the gastric epithelial cells from damage caused by luminal acid (2).

A highly acidic gastric content is directed toward the duodenum. To avoid ulceration, the duodenal mucosa like the gastric mucosa must protect itself from acidity. Similarly, a thickened mucus gel, increased blood flow, and a higher cellular buffering power protect the mucosa from injury. Through these integrated mechanisms, the epithelial cells are protected from damage due to repeated pulses of concentrated gastric acid (3).

Because of hiatal hernia, the gastric content may reflux toward the esophagus. This regurgitation may be asymptomatic or responsible for discomfort and pain.

The lumen of the normal esophagus is lined with nonsecreting epithelial cells. This repeated injury of the esophagus results in inflammation, architectural changes, and change in cell differentiation (Barrett's esophagus). Barrett's esophagus is a premalignant lesion, a consequence of reflux (4, 5).

Because of acidic reflux, the esophagus becomes lined with goblet cells rich in mucus, highlighted in biopsies using alcian blue pH 2.5 stain (4). Exposure to acid, by itself, contributes to the change of cell differentiation from squamous to glandular (5). It is therefore probable that the glandular differentiation of Barrett's esophagus and its complication, esophageal adenocarcinoma, are a consequence of the change in pH.

Inflammation and Loss of Cell Differentiation

A diagnosis of cancer cannot be established by examining a single cell suspension retrieved by needle biopsy or from a smear. It is by looking at the combination of the cellular (a cell is a few micrometers in diameter) and architectural abnormalities (a biopsy is a few millimeters in diameter) that the pathologist diagnoses the cancer. In clinical practice, a biopsy specimen comprising several thousand cells is necessary to ascertain the diagnosis.

There is a lack of reproducible difference between cells exposed to hypoxia during inflammation or in a cancer. In both cases, the cells display increased mitotic rate, change in morphology, as well as loss of original differentiation.

Increased FDG Uptake Correlates with Loss of Cell Differentiation

2-Deoxy-D-Glucose is a compound differing from glucose only in replacement of a hydroxyl group by a hydrogen atom. Marked by a radioactive isotope, it is preferentially bound by the tissues with high glycolytic metabolism which consumes greater quantities of glucose. Its use in trace quantities during positron emission tomography makes it possible to visualize anaerobic glycolysis (6–8).

Positron emission tomography is now routinely used to assess the extent of primary tumors as well as their metastases. There is a striking correlation between the loss of tumor cell differentiation and increased anaerobic glycolysis. There is increased uptake of radioactive 18F-flurodeoxyglucose (FDG), a direct consequence of glycolysis, in poorly differentiated cancer (9, 10).

The most aggressive tumor (higher tumor grade) usually displays a concomitant loss of cell differentiation, of cell-to-cell contact, and of tumor infiltration by lymphocytes. The uptake of 2-deoxy-D-glucose is more important in the cases of high-grade tumors than in the case of less aggressive tumors.

As in tumors, there is increased glycolysis during inflammation or subsequent fibrosis (6–8). These lesions can be imaged using positron emission tomography.

References

1. Van Seuningen I, Pigny P, Perrais M, Porchet N, Aubert JP (2001) Transcriptional regulation of the 11p15 mucin genes. Towards new biological tools in human therapy, in inflammatory diseases and cancer? Front Biosci 6:1216–1234
2. Engel E, Guth PH, Nishizaki Y, Kaunitz JD (1995) Barrier function of the gastric mucus gel. Am J Physiol 269:994–999
3. Kaunitz JD, Akiba Y (2002) Luminal acid elicits a protective duodenal mucosal response. Keio J Med 51:29–35
4. Van Eyken P (2000) Definition of Barrett's oesophagus. Acta Gastroenterol Belg 63:10–12
5. Fitzgerald RC, Omary MB, Triadafilopoulos G (1996) Dynamic effects of acid on Barrett's esophagus. An ex vivo proliferation and differentiation model. J Clin Invest 98:2120–2128
6. Zhuang H, Alavi A (2002) 18–fluorodeoxyglucose positron emission tomographic imaging in the detection and monitoring of infection and inflammation. Semin Nucl Med 32:47–59
7. Strauss LG (1996) Fluorine-18 deoxyglucose and false-positive results: a major problem in the diagnostics of oncological patients. Eur J Nucl Med 23:1409–1415
8. Wudel LJ Jr, Delbeke D, Morris D, Rice M, Washington MK, Shyr Y, Pinson CW, Chapman WC (2003) The role of 18F-fluorodeoxyglucose positron emission tomography imaging in the evaluation of hepatocellular carcinoma. Am Surg 69:117–124
9. Kubota K (2001) From tumor biology to clinical PET: a review of PET in oncology. Ann Nucl Med 15:471–486
10. Smith TA (1998) FDG uptake, tumor characteristics,and response to therapy a review. Nucl Med Commun 19:97–105

Response to Treatment: A Balance Between Glycolysis and Physical Constraints

CHAPTER 15

Cancer cell proliferation results from the combination of anaerobic glycolysis and lesser physical constraints. Conversely, cancer treatment inhibits cancer growth by depleting the ATP pool and increasing cellular constraints. After successful hormonal treatment for breast cancer, BCG for superficial bladder carcinoma, or after chemotherapy, the tumor is replaced by fibrosis. These increased extracellular constraints appear sufficient to explain tumor regression. Tumors growing in fibrotic tissues are unlikely to respond to increased fibrosis. Rapidly dividing tumors are surrounded by a loose stroma. These tumors are more likely to respond to chemotherapy.

Chemotherapy is mostly active in the treatment of malignancy among children and young adults. The extracellular matrix is softer in the young. Chemotherapy is more likely to increase these extracellular constraints and inhibit tumor growth.

Over five decades of intense research in cancer, drug discovery and development have provided about six dozen approved drugs for the treatment of malignancy (1). While major advances have been made in the treatment of children and young adults, the majority of cancer patients either fail to respond or will relapse and ultimately die from their metastatic disease.

Most Anticancer Drugs Have Been Discovered by Serendipity

Cancer is widely perceived as uncontrolled cellular growth. Accordingly, most antineoplastic agents share a common mechanism of action: they are cytotoxic. Anticancer agents are screened in vitro for their ability either to kill or to inhibit murine leukemia cells and tumor cell lines. More recently, molecular targets involved in cell cytotoxicity or growth inhibition have also been targeted. The screening process has evolved but the target has remained the same: the cancer cell.

This approach has yielded limited results and most drugs never made it to clinical care. There is a striking contrast between the "efficacy" of chemotherapy at killing cells grown on a Petri dish and the frequent lack of clinical results.

Cancer treatment cannot be summarized as cell kill. Cell lysis syndrome (i.e., the release in the blood stream of cellular components because of cell death) is frequent after infarction or trauma but rare after chemotherapy for solid tumor in adults, suggesting that chemotherapy does not immediately kill a great number of cancer cells. This is in sharp contrast with the unavoidable death of normal white blood cells. Within days after the inception of chemotherapy, the blood count starts to decrease. Similarly, it is within days after treatment that hair starts falling out. But tumor regression, when it occurs, has very different kinetics: it is much slower. Most treatment protocols call for tumor assessment 2–3 months after the start of treatment, but within days for measurement of blood count.

Chemotherapy and radiation therapy are highly effective at killing cells in vitro. But, in vivo, several arguments suggest that the main anticancer effect of chemotherapy may not be mediated by cell death.

One strong argument in favor of the extracellular effects of cancer treatment is the discrepancy between in vitro and in vivo data. For the treatment of infection, antibiotics are selected by exposing the bacteria to a panel of antibiotics. There is a highly reliable correlation between in vitro growth inhibition of the germ and successful treatment of the patient. Similarly, to select chemotherapy or radiation therapy and tailor the treatment, tumor cells are exposed in vitro to various anticancer chemicals. But the predictive value is poor (2). For example, Taghian evaluated the intrinsic radiation sensitivity of 58 cell lines derived from human malignant glioblastoma and found no significant correlation with the clinical outcome (3).

Breast cancer can be successfully treated by the antiestrogen tamoxifen. There is a contrast between the striking antitumor effect of tamoxifen in cancer patients and its lack of cell killing activity in vitro. MCF-7 is the foremost-established breast cancer cell line (4). Injected into athymic mice, MCF-7 cells grow and form tumors only in the presence of estrogen. Their growth is hormone dependent.

These MCF-7 cells are hormone independent in vitro. When growing on Petri dishes, the addition of hormone to the culture medium does not seem to affect the growth rate (4). The same discrepancy between in vitro and in vivo experiments is also found with other breast cancer cell lines (5).

In fact, the primary target of tamoxifen may be not the epithelial cells, but the breast itself. During treatment patients often notice breast swelling and hardening. Tamoxifen changes the architecture of the gland. It regulates the proliferation and the differentiation of fibroblasts, the attachment of the epithelial cells to fibroblasts, and the release of growth inhibitory proteins secreted by the fibroblasts which, in fact, target the cancer cells (4–10). A consequence of this extracellular effect of tamoxifen is the fact that after effective treatment the tumor is replaced by fibrosis.

Increased Constraints Explain Tumor Regression: The Example of Biphosphonate and BCG

Like tamoxifen in breast cancer, other medications are effective in treating cancer without been cytotoxic. They simply increase the physical constraints in the target organ (bladder for BCG, bone for biphosphonates). Accordingly, they are not effective in treating cancer in areas where they do not modify the extracellular matrix.

1) BCG Is Effective in Superficial but not in Metastatic Bladder Carcinoma

Intravesical administration of bacille Calmette-Guérin (BCG) eradicates 70% of carcinoma in situ for at least 1 year and prevents subsequent disease for 5 years in 60% of cases. In vitro, BCG has no direct cytotoxic effect (1, 11). Similar efficacy has been shown with mycobacterial cell wall extract (12) or other species of bacteria such as *Lactobacillus rhamnosus* (13) or even with cytotoxic chemotherapy. All these agents induce a nonspecific inflammation and fibrosis (11–14). BCG is responsible for local inflammation, with vesical irritability and pain at urination (1). BCG induces an infiltration of the bladder by mononuclear cells during the inflammatory reaction, which contributes to fibrosis of the bladder. This fibrosis may require cystectomy (15). One more argument in favor of the local effect is that BCG is effective only when delivered at the exact tumor site; BCG has no effect in the treatment of metastatic bladder carcinoma (1).

2) Biphosphonates Are Effective in Treating Bone Metastasis

Biphosphonates were synthesized and developed to treat osteoporosis. Biphosphonates have no major cytotoxic effects but change the architecture of the bone by inhibiting osteoclast-mediated bone resorption (16, 17). Accordingly, biphosphonates stop bone resorption from any cause, either osteoporosis or bone metastases.

The presence of tumor in bone is associated with activation of osteoclasts, resulting in excessive bone resorption and subsequent osteolysis (16). To the surprise of early researchers, biphosphonates prevent the occurrence and development of bone metastases.

In animals treated with biphosphonates, the subsequent injection of tumor cells failed to establish colonies in the bone (18). Similarly, in patients, biphosphonates reduced tumor growth and the occurrence of new bone metastases (19). The effect of biphosphonates is confined to the extracellular matrix of the bone. Biphosphonates have no palliative effect on metastases other than in the bone. Therefore, biphosphonate-associated tumor reduction in bone is most likely to be mediated by increased extracellular constraints.

The kinetics of tumor response after BCG or diphosphonate treatment is similar to response after cytotoxic chemotherapy. It takes a few weeks at the very least to assess the response to treatment, not hours or days.

Cytotoxic Chemotherapy Targets the Mitochondria

Apoptosis was long confined to tree specialists. It described the falling of leaves. It was long assumed that the genome of the tree would command, in the autumn, the change of color and death of the leaves. Today it is known that the leaves die of starvation because of shortened daylight and decreased energy resources.

Similarly, cytotoxic chemotherapy and radiation therapies are supposed to act through the induction of "apoptotic" cell death. Apoptosis can be activated through several different pathways, but these all appear to converge at a single event: mitochondrial membrane permeabilization. This "point-of-no-return" in cell death is the irreversible injury of the mitochondrial membrane (20, 21). Following successful chemotherapy, ATP necessary for cell survival is not available. The cell dies because of energy shortage.

Most cytotoxic drugs have been devised to target the DNA. Animal or vegetable cells have exactly the same DNA material in common (human DNA segments can be cultivated in bacteria). Similarly, the

DNA present in mitochondria has about the same chemical and three-dimensional structure as nuclear DNA; however, it is more fragile, because it is lacking in most repair systems (22). By targeting DNA, radiation therapy and cytotoxic chemotherapy target the mitochondria first.

Following injury, mitochondrial DNA repairs poorly (22). Energy is necessary for the survival of the cell and is provided by the mitochondria, which degrade glucose. Apoptosis may simply be a consequence of the failure of the mitochondria to provide adequate fuel for the cell.

Adriamycin is one of the major antitumor agents used for the clinical treatment of a wide variety of human cancers. In vitro, as early as 2 h after treatment with Adriamycin, the mitochondria undergo evident morphological changes. First, the mitochondria swell, then decrease in size and become picnotic. These changes precede cell death (23). Cisplatin or Methylgag exposures also debilitate mitochondrial functions (24, 25). Paclitaxel is supposed to be an antimicrotubule agent that induces mitotic block and apoptosis. Paclitaxel acts directly on mitochondria isolated from human cancer cells and induces the release of cytochrome C, one of the mitochondrial enzymes involved in energy transfer (26).

Chemotherapy Suppresses Respiration and Thus Induces Inflammation and Fibrosis

Radiation therapy and chemotherapy injure the mitochondria. Respiration is impaired. If the cell survives this injury, its metabolism switches to glycolysis. Inflammation is the stigma of transient glycolysis.

It is well known by clinicians that radiotherapy can cause inflammation. After a course of external beam radiotherapy, there is erythema of the normal skin. Cytotoxic chemotherapy also causes inflammation. Every single cytotoxic drug can induce inflammation and subsequent fibrosis. In hamsters, administration of bleomycin to the intrapleural cavity induces acute and chronic histological changes. At 7 days after injection of bleomycin, the pleura are thickened with infiltration of inflammatory cells, and formation of pleural fibrosis is seen at 28 days (30, 31). Similarly, the tyrosine kinase inhibitor STI571 (Gleevec), a promising treatment for chronic myeloid leukemia, is responsible for bone marrow fibrosis and reductions in cellularity (32).

In the days following effective chemotherapy, the tumor bed is infiltrated by lymphocytes (33, 34). This tumor infiltration and change in architecture appears to be a favorable prognostic predictor both for the short and the long term.

When chemotherapy is effective, the tumor may regress completely or more often partially. Extracellular matrix and fibrosis then replace the tumor. For example, 5-FU is used for treating liver metastasis from colorectal cancer. After preoperative chemotherapy, histological examination of the resected tumor shows marked degeneration, necrosis, fibrosis, and calcification (35). The same appears to be true for a wide variety of tumor sites (33, 36–38).

The organ microenvironment influences the response of metastases to chemotherapy. It is not uncommon to observe the regression of cancer metastases in one organ and their continued growth in other sites after systemic chemotherapy. These differences in response to treatment between the same tumors at different sites are not due to variations in the biodistribution of the chemotherapy (39).

Furthermore, not all cancer cells are killed by cytotoxic chemotherapy. Successful treatment of germ cell cancer and neuroblastoma is occasionally associated with residual tumors that on biopsy reveal only differentiated cells surrounded by fibrosis (37, 38).

Most long-term side effects result from this injuring of the respiration and subsequent deposition of extracellular matrix (27, 28). The use of Adriamycin in cancer chemotherapy has been limited due to its cumulative cardiovascular toxicity. The heart mitochondria represent almost 40% of the weight of the heart muscle.

Adriamycin interacts with mitochondrial cytochrome C oxidase and suppresses its enzyme activity. This lesion appears responsible for subsequent fibrosis and cardiac failure (29).Cisplatin is responsible for nephrotoxicity, nitrosourea or bleomycin is responsible for lung fibrosis. These side effects appear as a consequence of inflammation and fibrosis (40, 41). This inflammation and consequent fibrosis may, by itself, be the reason for secondary cancer (42).

Chemoresistance: Bypassing the Need for Cellular Respiration

Radiobiologists have pointed out for over 50 years that hypoxia or its consequence glycolysis decreases tumor radiosensitivity. When using ionizing radiation, the dose required to produce the same amount of cell killing is up to 3 times higher for hypoxic cells (43).

Multiple studies have shown better tumor control in patients with higher hemoglobin levels than in patients with levels below normal ranges due to tumor associated or therapy-induced anemia (44, 45). Lower hemoglobin levels result in decreased oxygen-carrying capacity and increased tumor hypoxia and radiation resistance (44).

Hypoxic cells are also resistant to chemotherapy. Tumor cells exposed to drugs immediately after transient hypoxia developed resistance to chemotherapy (46, 47). The expression of MDR1 (multidrug resistance 1), a transmembrane protein associated with tumor resistance to chemotherapeutics, is induced by ambient hypoxia. It is a consequence of the shift from respiration to glycolysis. There is an approximately sevenfold increase in MDR1 in epithelial cells exposed to hypoxia. Examination of the MDR1 gene identified a binding site for hypoxia-inducible factor-1 (HIF-1), and inhibition of HIF-1 expression by antisense oligonucleotides resulted in significant inhibition of hypoxia-inducible MDR1 expression and a nearly complete loss of basal MDR1 expression (47).

Resistance to treatment results from increased glycolysis. Glycolysis will be predominant source of energy. Treatment targeting the mitochondria will be ineffective, even counterproductive.

Conclusion

Cytotoxic chemotherapy targets cellular respiration. Warburg stated 50 years ago: "When one irradiates a tissue that contains cancer cells as well as normal cells, the respiration of the cancer cells, already too small, will decline further. If the respiration falls below a certain minimum that the cells need unconditionally, despite their increased fermentation, they die; whereas the normal cells, where respiration may be harmed by the same amount, will survive because, with a greater initial respiration, they will still possess a higher residual respiration after irradiation. This explains the selective killing action of X-rays on cancer cells."

In fact respiration and glycolysis are intertwined phenomenon. To function glycolysis needs NAD^+. Most of this NAD^+ is generated by the mitochondrial shuttles. The lesion of the mitochondria by conventional chemotherapy may result in a concomitant inhibition of glycolysis.

References

1. Grever M, Chabner B (1997) Cancer drug discovery and development in cancer. In: De Vita V, Hellman S, Rosenberg SA (ed) Principles and practice of oncology. Lippincott, Philadelphia, USA, pp 385–395
2. De Vita V (1997) Principles of cancer management: chemotherapy.In: De Vita V, Hellman S, Rosenberg SA (ed) Principles and practice of oncology. Lippincott, Philadelphia, USA
3. Taghian A, Ramsay J, Allalunis-Turner J, Budach W, Gioioso D, Pardo F, Okunieff P, Bleehen N, Urtasun R, Suit H (1993) Intrinsic radiation sensitivity may not be the major determinant of the poor clinical outcome of glioblastoma multiforme. Int J Radiat Oncol Biol Phys 25:243–249

4. Sonnenschein C, Soto AM (1999) The society of cells. Bios Scientific Publishers, Oxford, UK
5. Millon R, Nicora F, Muller D, Eber M, Klein-Soyer C, Abecassis (1989) Modulation of human breast cancer cell adhesion by estrogens and antiestrogens. J Clin Exp Metastasis 7:405–415
6. Colletta AA, Wakefield LM, Howell FV, Van Roozendaal KE, Danielpoor D, Ebbs SR, Sporn MB, Baum M (1990) Anti-estrogens induce the secretion of active transforming growth factor beta from human fetal fibroblasts. Br J Cancer 62:405–409
7. Schwartz L (1983) Le mRNA Ps2 dans le cancer du sein. Thèse de médecine, Strasbourg
8. Rajah TT, Rambo DJ, Dmytryk JJ, Pento JT (2001) Influence of antiestrogens on NIH-3T3-fibroblast-induced motility of breast cancer cells. Chemotherapy 47:56–69
9. Woodward TL, Mienaltowski AS, Modi RR, Bennett JM, Haslam SZ (2001) Fibronectin and the alpha(5)beta(1) integrin are under developmental and ovarian steroid regulation in the normal mouse mammary gland. Endocrinology 142:3214–3222
10. Branham WS, Lyn-Cook BD, Andrews A, Sheehan DM (1993) Growth of separated and recombined neonatal rat uterine luminal epithelium and stroma on extracellular matrix. Cell Dev Biol Anim 29:408–414
11. Kamat AM, Lamm DL (2001) Immunotherapy for bladder cancer. Curr Urol Rep 2:62–69
12. Morales A, Chin JL, Ramsey EW (2001) Mycobacterial cell wall extract for treatment of carcinoma in situ of the bladder. J Urol 166:1633–1637
13. Lim BK, Mahendran R, Lee YK, Bay BH (2002) Chemopreventive effect of Lactobacillus rhamnosus on growth of a subcutaneously implanted bladder cancer cell line in the mouse. Jpn J Cancer Res 93:36–41
14. Marshall BG, Wangoo A, Cook HT, Shaw RJ (1996) Increased inflammatory cytokines and new collagen formation in cutaneous tuberculosis and sarcoidosis. Thorax 51:1253–1261
15. Torrubia Romero FJ, Martinez Rodriguez J, Cruz Navarro N, Ramirez Mendoza A, Soltero Gonzalez A (1996) Complications of immunotherapy with BCG. Report of a case with retractile bladder. Arch Esp Urol 49:74–78
16. Mundy GR, Yoneda T, Higara T (2001) Preclinical studies with zoledronic acid and other biphosphonates: impact on the bone environment. Semin Oncol 28:35–44
17. Healy JH. Metastatic cancer to the bone in Cancer. In: De Vita V, Hellman S, Rosenberg SA (ed) Principles and practice of oncology. Lippincott, Philadelphia, USA
18. Tamura H, Ishii S, Enomoto K (1994) Evaluation of the efficacy of biphosphonates to bone metastasis of breast cancers in a rat model. Proc Ann Am Assoc Cancer Res 35:A2787

19. Marston Linehan W, Cordon-Cardo C, Isaacs W (1997) Cancers of the genitourinary systems in Cancer. In: De Vita V, Hellman S, Rosenberg SA (ed) Principles and practice of oncology. Lippincott, Philadelphia, USA
20. Marchetti P, Mortier L, Beauvillain V, Formstecher P (2002) Are mitochondria targets of anticancer drugs responsible for apoptosis? Ann Biol Clin (Paris) 60:391–403
21. Debatin KM, Poncet D, Kroemer G (2002) Chemotherapy: targeting the mitochondrial cell death pathway. Oncogene 21:8786–8803
22. Mandavilli BS, Santos JH, Van Houten B (2002) Mitochondrial DNA repair and aging. Mutat Res 509:127–151
23. Serafino A, Sinibaldi-Vallebona P, Lazzarino G, Tavazzi B, Di Pierro D, Rasi G, Ravagnan G (2000) Modifications of mitochondria in human tumor cells during anthracycline-induced apoptosis. Anticancer Res 20:3383–3394
24. Dorward AM, Singh G (1996) Energetic characteristics of cisplatin resistant ovarian carcinoma cells. Anticancer Res 16:443–447
25. Pathak SN, Porter CW, Dave C (1977) Morphological evidence for an antimitochondrial action by methylglyoxal-bis(guanylhydrazone). Cancer Res 37: 2246–2250
26. Andre N, Braguer D, Brasseur G, Goncalves A, Lemesle-Meunier D, Guise S, Jordan MA, Briand C (2000) Paclitaxel induces release of cytochrome c from mitochondria isolated from human neuroblastoma cells. Cancer Res 60: 5349–5353
27. Breuer K, Diehl V, Ruffer U (2000) Long-term toxic sequelae of the treatment of Hodgkin's disease. Med Klin 95:378–384
28. Sommer K, Brockmann WP, Hubener KH (1990) Treatment results and acute and late toxicity of radiation therapy for testicular seminoma. Cancer 66: 259–263
29. Papadopoulou LC, Theophilidis G, Thomopoulos GN, Tsiftsoglou AS (1999) Structural and functional impairment of mitochondria in adriamycin-induced cardiomyopathy in mice: suppression of cytochrome c oxidase II gene expression. Biochem Pharmacol 57:481–489
30. Yamaguchi E, Nagai A, Sakamoto K, Takizawa T (1991) Morphological changes in the pleura caused by intracavitary administration of bleomycin. Gan To Kagaku Ryoho 18:971–975
31. Hay J, Shahzeidi S, Laurent G (1991) Mechanisms of bleomycin-induced lung damage. Arch Toxicol 65:81–94
32. Hasserjian RP, Boecklin F, Parker S, Chase A, Dhar S, Zaiac M, Olavarria E, Lampert I, Henry K, Apperley JF, Goldman JM (2002) ST1571 reduces bone marrow cellularity and normalizes morphologic features irrespective of cytogenetic response. Am J Clin Pathol 117:360–367
33. Demaria S, Volm MD, Shapiro RL, Yee HT, Oratz R, Formenti SC, Muggia F, Symmans WF (2001) Development of tumor-infiltrating lymphocytes in breast cancer after neoadjuvant paclitaxel chemotherapy. Clin Cancer Res 7: 3025–3030

34. Magnarin M, Bergamo A, Carotenuto ME, Zorzet S, Sava G (2000) Increase of tumour infiltrating lymphocytes in mice treated with antimetastatic doses of NAMI-A. Anticancer Res 20:2939–2944
35. Takeda Y, Hasuike Y, Mishima H, Nishisho I, Kikkawa N (2000) Case report–efficacy of short-term intraarterial 5-fluorouracil for liver metastasis from colorectal cancer. Gan To Kagaku Ryoho 27:1838–1841
36. Yonemura Y, Kinoshita K, Fujimura T, Fushida S, Sawa T, Matsuki N, Tanaka S, Kamata T, Takashima T, Miyazaki I (1996) Correlation of the histological effects and survival after neoadjuvant chemotherapy on gastric cancer patients. Hepatogastroenterology 43:1260–1272
37. Hain SF, O'Doherty MJ, Timothy AR, Leslie MD, Harper PG, Huddart (2000) Fluorodeoxyglucose positron emission tomography in the evaluation of germ cell tumours at relapse. Br J Cancer 83:863–869
38. Matsuyama H, Yamamoto N, Sakatoku J, Suga A, Hayashida S, Kamiryo Y, Nagata K, Naito K (1994) Predictive factors for the histologic nature of residual tumor mass after chemotherapy in patients with advanced testicular cancer. Urology 44:392–398
39. Fidler IJ, Wilmanns C, Staroselsky A, Radinsky R, Dong Z, Fan D (1994) Modulation of tumor response to chemotherapy by the organ environment. Cancer Metastasis Rev 13:209–222
40. Hartmann JT, Kollmannsberger C, Kanz L, Bokemeyer C (1999) Platinum organ toxicity and possible prevention in patients with testicular cancer. Int J Cancer 83:866–869
41. Massin F, Coudert B, Foucher P, Lombard JN, Reybet-Degat O, Jeannin L, Camus P (1992) Nitrosourea-induced lung diseases. Rev Mal Respir 9:575–582
42. L. Schwartz L, Balosso J, Baillet F, Brun B, Amann JP, Sasco A (2002) Cancer, the role of extracellular disease: an hypothesis. Med Hypothesis 58:340–346.
43. Gray LH, Conger AD, Ebert M, Hornsey S, Scott OC (1953) Concentration of oxygen dissolved in tissue at time of irradiation as a factor in radiotherapy. Br J Radiol 26:638–648
44. Thomas GM (2002) Raising hemoglobin: an opportunity for increasing survival? Oncology 63S:19–28
45. Feldmann HJ (2001) Oxygenation of human tumors – implications for combined therapy. Lung Cancer 33:S77–83
46. Sakata K, Kwok TT, Murphy BJ, Laderoute KR, Gordon GR, Sutherland RM (1991) Hypoxia-induced drug resistance: comparison to P-glycoprotein-associated drug resistance. Br J Cancer 64:809–814
47. Comerford KM, Wallace TJ, Karhausen J, Louis NA, Montalto MC, Colgan SP (2002) Hypoxia-inducible factor-1-dependent regulation of the multidrug resistance (MDR1) gene. Cancer Res 62:3387–3394
48. Hue L, Rousseau GG (1993) Fructose 2,6-bisphosphate and the control of glycolysis by growth factors, tumor promoters and oncogenes. Adv Enzyme Regul 33:97–110

Targeting Glycolysis

CHAPTER 16

PET scans demonstrate an increased uptake of 2-deoxy-d-Glucose by human cancers. Preliminary in vitro and in vivo studies scattered in the literature of the past 50 years suggest an inhibitory effect of the glucose antagonist, 2-deoxy-d-Glucose. This molecule has been given orally, at the same dose to mice, in order to mimic caloric restriction without major side effect. As of today, no human clinical trials questioning the validity of inhibition of cancer growth with 2-deoxy-d-Glucose have been reported.

Until the end of the twentieth century, it was firmly believed that a living cell cannot change its destiny, cannot differentiate into a cell of another genetic line (1). This belief resembles the notion of predetermination. During the Middle Ages, some religious teachers held that our destiny is determined before our birth. For their own sake, heretics had to be killed.

For man, like for the cell, such a belief ignores the weight of the environment. As a man, I could have had a different destiny. Had my environment been slightly different, I could have ended up as a farmer in Alsace, this region of France that I love so dearly. I am not a theologian, but I know that as far as the cell is concerned this belief of predestination has long been contradicted in clinical practice. There are other options than killing the "malignant" cell.

In childhood, during growth, cartilage becomes ossified. But even in old age, after a fracture, bone cells "reverse" into cartilage during the process of bone healing. More recently, it was organ transplants that put a definite end to the notion of definitive cell differentiation. A woman's heart is transplanted into a man. The cells of the patient colonize the grafted organ and change into cardiac cells (2). Hematopoietic cells can differentiate into liver or even lung cells (1).

The destiny of cancer cells is no more rigidly fixed than that of normal cells. Like the normal cell, the cancer cell is adaptable. A cancer cell injected into an embryo, or directly into the liver of a young animal, looses its malignancy and changes into an ordinary cell exactly like its neighbors which originated from healthy parent cells (3).

Inhibiting tumor growth is perhaps simply a matter of inhibiting glycolysis without blocking cell respiration and, therefore, without killing the cell. This idea was hindered for a long time by a tenet, which came from Warburg himself. But Warburg was wrong when he said, "If an injury of respiration is to produce cancer, this injury must be irreversible." We understand by this not only that the inhibition of respiration remains after removal of the respiratory poison but, even more, that the inhibition of respiration also continues through all the following cell divisions, for measurements of metabolism in transplanted tumors have shown that cancer cells cannot regain normal respiration, even in the course of many decades, once they have lost it (4).

Catabolic Repression

In order for yeast to ferment and for sugar to be partly burned into alcohol, the brewer adds great quantities of malt (i.e., sugar) to the hops. The presence of sugar causes partial inhibition in the respiration of the yeast, resulting in the fermentation and the synthetization of a product of partial glucose combustion: alcohol. The Crabtree effect (named after its discoverer) is reversible; as soon as concentrations become low in glucose, respiration goes back to normal: glycolysis is inhibited (5). High glucose concentrations or low oxygen content inhibits the respiration of the human cell, like that of yeast (6, 7).

Yeast encloses many types of energy flux, often in competition. Blocking glycolysis by glucose withdrawal or by oxygen input will result in predominant respiration. Inversely, diminishing oxygen will result in predominant glycolysis. The addition of micromolar concentrations of glucose (similar to the concentration in the human serum) to the culture medium results in the Crabtree effect, a decrease of cell respiration rate and a concomitant increase in glycolysis (6).

Like brewer's or baker's yeast, human cells, normal or cancerous, can either breathe or obtain energy through fermentation. In a human cell, glycolysis is inhibited in the presence of oxygen or low glucose concentration. This effect is not limited to normal cells but has been repeatedly observed in malignant cells (6, 7).

Just like in yeast, tumor glycolysis is adjustable. Proof ad absurdum, mathematicians would say. Cancer is easy to cure in a Petri dish. Cells are cultured in an incubator, in an oxygen-enriched milieu. High oxygen concentration facilitates respiration and inhibits glycolysis (8, 9). Chemotherapy, radiotherapy, and hormone therapies are all more effective when the cell is in an aerobic rather than a hypoxic milieu.

It is easy to restore cell respiration in a Petri dish with the help of oxygen, because the oxygen is easily diffused through the cell microns. In the human body, like in the ponds of Alsace, oxygen is poorly diffused. Restoring cell respiration in a cancer patient is very difficult because the distance between the capillary and the tumor cell is often too great. Oxygenation using a hyperbaric chamber has proved to be of little clinical benefit. Hyperbaric oxygen therapy as a revolutionary technique had to be abandoned.

Chemical Inhibition of Glycolysis

Let us state the obvious: there is no glycolysis without glucose. Over a few dozen years, articles were published correlating the effect of glucose depletion and tumor inhibition (10–13). Sugar depletion has little effect on normal cells (they can use other nutrients), but it produces reproductive inhibition in even the most aggressive tumor cells. Fasting resulted in a linear dose-response inhibition of tumor growth (12).

Depriving a cancer patient of sugar has little chance of curing him. The liver is rich in glycogen and, in case of depletion in glucose, the liver degrades its stores of glycogen into glucose and pours it into the blood. Therefore, medication is needed to chemically inhibit glycolysis inside the tumor cell.

Let us go back to notions of basic chemistry acquired in first year medical school and buried far back in our memory. Glucose is metabolized into pyruvate through the process of glycolysis. Energy output is low: two ATP molecules per molecule of glucose. Pyruvate can be either broken down into lactic acid or metabolized in the Krebs cycle: in that case, cell respiration reduces pyruvate to carbonic gas and water, releasing larger amounts of ATP. A reversible enzyme, lactate dehydrogenase, controls the transfer of pyruvate into lactic acid. During muscular exertion, oxygen supply is insufficient and glycolysis predominates, resulting in the synthesis of lactic acid. At rest, oxygen supply increases and lactic acid is further metabolized into pyruvate at first and then broken down during the Krebs cycle into carbonic gas, water, and more energy.

Like glucose, lactic acid is a cellular fuel. In case of distress, such as septic shock, Ringer solution rich in lactic acid is transfused. Because lactate dehydrogenase is a reversible enzyme, lactic acid will be rapidly transformed into energy producing metabolized pyruvate in the Krebs cycle.

Chemists and pharmacologists have long studied glycolysis inhibitors. But very few studies are being conducted on the promising properties of these agents. The focus is no longer on cancer as fermentation of living matter, but rather on the details of cell mechanics. Nevertheless, it is likely that glycolysis inhibition would be sufficient to achieve cancer regression through the combined effect of reduced glycolysis and persistent constraint from the stroma.

Reversing the Energetic Flux

Lactate dehydrogenase is a key enzyme which converts pyruvate into lactic acid. This step is reversible. If there is an excess of pyruvate like in glycolysis, there will be synthesis of large concentrations of lactic acid. If there is an excess of lactic acid, that molecule will be transformed back into pyruvate and metabolized into the Krebs cycle. An input of lactic acid inhibits glycolysis and thus stimulates cell respiration. Thus, the intake of dietary supplements of mixed cultures of several lactic acid bacteria inhibits the early development of colon adenomas; the inhibition of microadenomas results in a reduction of subsequent polyp and tumor yield in the mouse colon (14).

Molecules similar to lactic acid, usually cyclic polylactates used as decoys, inhibit lactate dehydrogenase and cell growth (15, 16). Lactic acid synthesis and, consequently, anaerobic glycolysis are inhibited. About half of the anaerobic glycolytic activity of FM3A ascite tumor cells was inhibited, and tumor cell growth was also effectively inhibited by cyclic polylactates. Mice treated with cyclic polylactates after inoculation of FM3A ascite tumor cells lived significantly longer than mice not treated with cyclic polylactates (16). Similarly, a lactate transport inhibitor, guercetin, inhibits lactate release and results in tumor cell acidification (17). Combined treatment with doxorubicin and a lactate dehydrogenase gives better results than additive cell killing (18).

2-Deoxy-d-Glucose: The First candidate for Clinical Trials

2-Deoxy-D-glucose is a compound differing from glucose only in the replacement of a hydroxyl group by a hydrogen atom. 2-Deoxy-D-glucose has been employed for a variety of purposes, ranging from tracing native glucose molecule transport and processing to actually blocking energy metabolism. Marked by a radioactive isotope, it is preferentially bound by the tissues with a high glycolytic metabolism, which consume greater quantities of glucose. Its use in trace quantities during tomography makes it possible to visualize anaerobic glycolysis, and therefore inflammation, fibrosis, and cancer (19–21).

Glucose binds to the enzyme phosphohexose isomerase, which converts glucose-6-phosphate into fructose-6-phosphate. The primary effect of 2-deoxy-D-glucose is to serve as an extremely strong competitor of glucose. As a consequence, the concentration of 2-deoxy-D-glucose-6-phosphate builds up and is eventually excreted, while metabolism of native glucose is greatly reduced.

In vitro, the addition of 2-deoxy-D-glucose to the culture media inhibits cancer growth (23–29). Kern tested the ability of 2-deoxy-D-glucose to inhibit the growth of an established methylcholanthrene-induced rat fibrosarcoma in F344 rats (24). Rats were randomized to receive intraperitoneally administered 2-deoxy-D-glucose or saline solution at doses of 0.75 gm/kg, or 1.75 gm/kg, beginning 3 days after tumor implantation. After 10 days of treatment, the rats treated with 2-deoxy-D-glucose had a significant reduction in tumor weight: 50%–70% when compared with controls (24). Lower doses appear as effective as high doses. Fifteen minutes after injection the rats appeared to be in a transient state of lethargy for a few hours but recovered quickly. There was no detectable long-term toxicity in the two lower dose groups.

Similarly, continuous infusion of 2-deoxy-D-glucose (400 mg/kg) into the arterial artery significantly reduced the tumor burden of colonic metastases in male Wistar rats (25). Karczmar (26) also confirmed that 2-deoxy-D-glucose (2 g/kg) inhibits glycolysis in rats suffering from large sarcoma. This dosage had no noticeable side effects.

Inhibition of glycolysis with 2-deoxy-D-glucose makes tumor cells more sensitive to injury due to cell respiration by TNF (27), cytotoxic agents (28), or radiation therapy (29).

The long-term toxicity of 2-deoxy-D-glucose appears limited. The data come from a research area apparently far removed from oncology: aging. Since 1935, laboratory experiments have repeatedly demonstrated that restricting the number of calories consumed increases both mean and maximal life span in numerous creatures. This has been shown in worms, water fleas, flies, spiders, guppies, rodents, and most recently in nonhuman primates: Rhesus monkeys. Animals whose calorie intake is restricted by 30%–40% live 30%–40% longer. They demonstrate a later and lower incidence of many age-related diseases such as cardiovascular disease or cancer. The calorie-restricted animals experience decreased glycolysis which results in lower body mass, slower growth, lower temperature, and lower secretion of by-products of glucose, such as hormones resulting in delayed puberty. The long-term toxicity of 2-deoxy-D-glucose is similar to severe dieting. Treatment with 2-deoxy-D-glucose has also been used to mimic dieting. Partial replacement of glucose by 2-deoxy-D-glucose reduces glucose/energy flux without decreasing food intake in rats. In a 6-month pilot study in rats, 2-deoxy-D-glucose lowered plasma insulin and body temperature in a manner similar to that of caloric restriction (30–32).

To my knowledge, there are almost no available data on humans treated for cancer with 2-deoxy-D-glucose. The reason is simple. 2-Deoxy-D-glucose is a natural molecule. Its chemical synthesis has been known for a long time. 2-Deoxy-D-glucose cannot be patented and is hardly a source of business and profit for industry.

The quest for the miracle drug is not confined to the data banks of the medical literature. One dream, which might become reality, is to treat cancer successfully. Miracles do occur.

Popular wisdom holds that in order to succeed you need one-third knowledge, one-third know-how, and one-third "ability-to-let-others-know." Our ambition was to meet all three of these conditions.

References

1. Fagot-Largeault A (2002) Implications éthiques sur la recherche sur les cellules souches humaines et le transfert nucléaires. Philosophie de sciences éthiques. Colloque francodanois. Fondation Paule Mikkelsen
2. Bayes-Genis A, Salido M, Sole Ristol F, Puig M,Brossa V, Camprecios M, Corominas JM, Marinoso ML, Baro T, Vela MC, Serrano S, Padro JM, Bayes de Luna A, Cinca J (2002) Host cell-derived cardiomyocytes in sex-mismatch cardiac allografts. Cardiovasc Res 56 : 404–410
3. Schwartz L, Balosso J, Baillet F, Brun B, Amann JP, Sasco A (2002) Cancer, the role of extracellular disease: an hypothesis. Med Hypothesis 58 : 340–346
4. Warburg O (1956) "On The Origin of Cancer Cells." Science 123 : 309–314
5. Aon MA, Cortassa S (1998) Catabolite repression mutants of Saccharomyces cervisiae. Biotechnol Bioeng 59 : 203–213
6. Wojtczack L, Teplova V, Bogucka K, Czyz A, Makowska A, Wieckowski M, Duszynski J, Evtodienko Y (1999) Effect of glucoe and desoxyglucose on the redistribution of calcium in Ehrlich ascites tumor. Eur J Biochem 263 : 495–501
7. Rodriguez-Enriquez S, Juarez O, Rodriguez-Zavala J, Moreno-Sanchez R (2001) Multisites control of the Crabtree effect in hepatoma cells. Eur J Biochem 268 : 2512–2519
8. Guppy M, Leedman P, Xu X, Russell V (2002) Contribution by different fuels and metabolic pathways to the total ATP turnover of proliferating MCF-7 breast cancer cells. Biochem J 364 : 309–315
9. Nielsen FU, Daugaard P, Bentzen L, Stodkilde-Jorgensen H, Overgaard J, Horsman MR, Maxwell RJ (2001) Effect of changing tumor oxygenation on glycolytic metabolism in a murine C3H mammary carcinoma assessed by *in vivo* nuclear magnetic resonance spectroscopy. Cancer Res 61 : 5318–5325
10. Hue L, Rousseau GG (1993) Fructose 2,6-bisphosphate and the control of glycolysis by growth factors, tumor promoters and oncogenes. Adv Enzyme Regul 33 : 97–110
11. Pauwels PJ, Opperdoes FR, Trouet A (1985) Effects of antimycin, glucose deprivation, and serum on cultures of neurons, astrocytes, and neuroblastoma cells. J Neurochem 44 : 143–148
12. Sandor RS (1976) Effects of fasting on growth and glycolysis of the Ehrlich ascites tumor. J Natl Cancer Inst 56 : 427–428
13. Singh G, Lakkis CL, Laucirica R, Epner DE (1999) Regulation of prostate cancer cell division by glucose. J Cell Physiol 180 : 431–438

14. Fukui M, Fujino T, Tsutsui K, Maruyama T, Yoshimura H, Shinohara T, Fukui M, Nada O (2001) The tumor-preventing effect of a mixture of several lactic acid bacteria on 1,2-dimethylhydrazine-induced colon carcinogenesis in mice. Oncol Rep 8:1073–1078
15. Aizawa S, Shimizu N, Handa H, Hiranda M, Hoshi H, Nagasu M, Kanno H, Nagasu Y, Imanishi Y (2000) Effects of cyclic polylactate on the growth of cloned leukemic cells *in vitro*. Hematol Oncol 18:51–60
16. Takada S, Nagato Y, Yamamura M (1997) Effect of cyclic polylactates on tumor cells and tumor bearing mice. Biochem Mol Biol Int 43:9–17
17. Volk C, Kempski B, Kempski OS (1997) Inhibition of lactate export by quercetin acidifies rat glial cells *in vitro*. Neurosci Lett 223:121–124
18. Hamilton E, Fennell M, Stafford DM (1995) Modification of tumour glucose metabolism for therapeutic benefit. Acta Oncol 34:429–33
19. Strauss LG (1996) Flurine-18 deoxyglucose and false-positive results: a major problem in the diagnostics of oncological patients. Eur J Nucl Med 23: 1409–1415
20. Wudel LJ Jr, Delbeke D, Morris D, Rice M, Washington MK, Shyr Y, Pinson CW, Chapman WC (2003) The role of 18F-fluorodeoxyglucose positron emission tomography imaging in the evaluation of hepatocellular carcinoma. Am Surg 69:117–124
21. Kubota K (2001) From tumor biology to clinical PET: a review of PET in oncology. Ann Nucl Med 15:471–486
22. Griffiths M, Keast D, Patrick G, Crawford M, Palmer TN (1993) The role of glutamine and glucose analogues in metabolic inhibition of human myeloid leukaemia in vitro. Int J Biochem 25:1749–1755
23. Bentley J, Bell SE, Quinn DM, Kellett GL, WarrJR (1996) 2-deoxy-D-Glucose toxicity and transport in human multidrug-resistant KB carcinoma cell lines. Oncol Res 8:77–84
24. Kern KA, Norton JA (1987) Inhibition of established rat fibrosarcoma growth by the glucose antagonist 2-deoxy-D-Glucose. Surgery 102:380–385
25. Cay O, Radnell M, Jeppsson B, Ahren B, Bengmark S (1992) Inhibitory effect of 2-deoxy-D-Glucose on liver tumor growth in rats. Cancer Res 52:5794–5796
26. Karczmar GS, Arbeit JM, Toy BJ, Speder A, Weiner MW (1992) Selective depletion of tumor ATP by 2-deoxyglucose and insulin, detected by 31P magnetic resonance spectroscopy. Cancer Res 52:71–76
27. Halicka HD, Ardelt B, Li X, Melamed MM, Darzynkiewicz Z (1995) 2-deoxy-D-Glucose enhances sensitivity of human histiocytic lymphoma U937 cells to apoptosis induced by tumor necrosis factor. Cancer Res 55:444–449
28. Griffiths M, Keast D, Patrick G, Crawford M,Palmer TN (1993) The role of glutamine and glucose analogues in metabolic inhibition of human myeloid leukaemia in vitro. Int J Biochem 25:1749–1755
29. Jha B, Pohlit W (1993) Reversibility of inhibition of DNA double strand break repair by2-deoxy-D-Glucose in Ehrlich ascites tumour cells. Int J Radiat Biol 63:459–467

30. Lane MA, Black A, Handy A, Tilmont EM, Ingram DK, Roth GS (2001) Caloric restriction in primates. Ann N Y Acad Sci 928:287–295
31. Roth GS, Ingram DK, Lane MA (2001) Caloric restriction in primates and relevance to humans. Ann N Y Acad Sci 928:305–315
32. Lane MA, Baer DJ, Rumpler WV, Weindruch R, Ingram DK, Tilmont EM, Cutler RG, Roth GS (1996) Calorie restriction lowers body temperature in rhesus monkeys, consistent with a postulated anti-aging mechanism in rodents. Proc Natl Acad Sci USA 93:4159–4164

Conclusion

CHAPTER 17

One more book about cancer, you say. And it won't be the last, I might add. Because we must admit that, as for as cancer is concerned, we have not yet found a definitive solution, the miracle – no other word will do – which sometimes occurs on the tentative trails of science, and which would change this dreaded illness into a curable disease.

To be a doctor is to deal with daily realities that do not always conform to our hopes. And all our hopes are directed toward one and the same concrete reality: effective treatment. Of course, over the years, we have uncovered trails whose validity has been established. So much, the better. We have also discovered long-forgotten facts and have let them guide us. In short, we count ourselves among those travelers who do not expect to see the lights of Rome halfway along the journey.

But we continue. We are, after all, heirs of the scientific tradition, with its victories and errors. The earth is no longer the center of the universe, it is part of a galaxy which is, in turn, one among many. The image might be less glorious, but we see a larger picture and we know that the unimaginable is possible. We thought that our world was the only one possible, and we discovered that the world of possibilities is infinite. The history of the earth is marked by massive extinctions: that of the pre-Cambrian flora, that of the dinosaurs; but also by fabulous renewals: the human species, for one.

Evolution is a matter of external pressure, as Darwin rightly observed. Our cells are subjected to similar constraints. We do not know how, but they survive in a hostile milieu.

But hasn't this always been true of humanity as a whole? We wanted to believe that the behavior of our cells was governed by an orderly system which instructs white blood cells to combat infection, neurons to colonize the skull. We need the illusion of a logical system. Just like ant communities where each member carries out its specialized function, intestinal cells process food for the community, macrophages do the cleaning, and the immune system provides defense against aggression. This organization is efficient. We wanted to apply the same model to the notion of the genome. The genome is widely perceived as giving orders, stimulating lazy cells, slowing down those which are overactive.

A friend who is an immunologist explained to me one day that lymphocytes spread out over a Petri dish. He has two explanations for this phenomenon. The first involves adhesion molecules, their receptor, as well as the cytoskeleton. But the explanation he likes better is that lymphocytes are always hungry and spread out to eat the plastic!

Who has not marveled at the symbiosis between the shark and those little fish which feed on the debris of food between his teeth? No genetic program can explain this symbiosis, and that is precisely where the marvel lies! The cell's differentiation, that is, its metabolism, depends on what the cell can feed on and on constraints from its neighbors. So much for the genome as central commander!

The idea that the genome does not explain cancer began to be whispered by some rebellious professionals at first, and was lately reiterated more explicitly by unconventional scientists.

There were a number of inconsistencies: the ability of cancer cells to become normal again, the ability to produce cancer without lesions in the genome, and above all, the frequent absence of effective therapies. Answers to these questions have been suggested, but they are more and more like the epicycles invented by the ancients to prove that the sun revolves around the earth.

We looked for a cancer gene. We did not find it. Since we did not find one, we thought it would be smarter to look for several. Articles on multifactorial, polygenic illnesses flooded the literature. All of them contradicted by clinical practice! All liver cancers appear on an already damaged liver, all lung cancers on sick lungs.

As I write this, I am watching the animation of a Paris street. People speak to each other, exchange comments about the weather and information about the prices of produce, which have gone up again. At first sight, it might seem that these people have gathered together around a common cause, with a precise goal in mind. But in fact, as we know from experience, they are only there by chance and continue on their way, each one seeking his own selfish satisfaction. A bird's eye view of the animation of the neighborhood shows it to be made up of a succession of small personal errands undertaken for the most individualistic reasons.

Today, we have to decipher this new scene which includes the genome but no longer places it at the center.

Cells are individual entities. They try to survive in an often hostile environment. Cellular plasticity is nothing other than an illustration of this attempt to survive. Consequently, the genome can be seen as a giant tool box allowing the cells to deal with mechanical and biochemical stress. Mutations are then the result, rather than the cause of stress.

Let us reverse the paradigm by creating another scene whose focus is not the genetic stigma of the stress, but life itself. We are made of stardust. Life started in a world where oxygen was hard to come by. Photosynthesis produced by algae freed this nutrient. Today, we live in a world rich in oxygen. As available energy, oxygen allowed evolution toward multicellular, complex beings such as dinosaurs an mammals. Bacteria are simple organisms; they are immortal. They divide infinitely and are not condemned to die. But we are.

Life has different rhythms for different people. We say of someone that he looks older or younger than he is. Every doctor knows that someone who has aged quickly, no matter what his actual age, is exposed to a greater a risk of cancer. With cancer as with everything else, it is better to be rich and well educated than to be a manual laborer. In the latter category, premature aging plays a important part in excessive mortality rates. The worker pays the price for a more active metabolism, for greater oxygen consumption.

Emphasis on the role of cellular environment destroys another tenet: the irreversibility of cell differentiation. The "party line," as it were, was that a tissue which has become differentiated never reverses to change differentiation. But party loyalty requires that we be blind and deaf. The first refutation came from transplantation. The heart of a woman can be grafted to a male host. A few months later, male myocardial cells have partially colonized that female heart. Bone marrow cells have changed destiny.

Similarly, if you inject crushed glass into the pericardium, the latter will become ossified. If you decrease vascularization of a tendon, the tendon center will first be replaced by cartilage, then by bone, and finally by hematopoietic marrow.

Changing paradigms may have profound consequences. The first example came from the pediatricians: if there is congenital atrophy of the bile duct, the underlying choleic stasis produces fibrosis. When bile duct permeability is restored, choleic stasis is amended and the cirrhosis which was thought to be irreversible disappears. Thus, if the physical constraints are changed, cells can change differentiation and architectural disorder can regress.

Similarly, there is documented evidence that a change in chemical constraints can reverses fibrosis. In the months or years following radiotherapy, some patients develop first an inflammation in the irradiated region, then a painful patch. The tissue hardens, new vessels develop. At histology, there is an accumulation of extracellular matrix: we speak of radiation-induced fibrosis. These complications can follow either intentional radiotherapy to destroy a particularly aggressive tumor, or an error in radiation dosage. Contrary to hepatic fibrosis, radiation-induced fibrosis is a good animal model. Research in this field was extensive, thanks to military interest in the possible benefits of treatment for survivors of an atomic attack.

For the past 30 years, the Francophone medical literature has suggested that radiation-induced fibrosis is reversible. Two dozen publications have attested to the efficiency of super oxide dismutase (SOD). Why has this information not yet been put to use? First, SOD cannot be patented, its discovery dates back over 50 years. Therefore, the pharmaceutical industry cannot protect its research and make a profit. The second reason is that SOD used to depend on bovine extraction. Therefore, its production was prohibited during the "mad cow disease" crisis. The SOD marketed today is genetically produced or made from vegetable sources. The major market for this molecule is the treatment of another architectural disorder: wrinkles! The cosmetics industry does not want to frighten its clients by linking the use of this product to cancer prevention.

Experts are studying action strategies. As for me, I have seen two patients transformed by this treatment. After the treatment, the new vessels had regressed and fibrosis was reduced or eliminated. This treatment offers the greatest benefit to patients with radiation-induced fibrosis.

Under the microscope, all fibroses, whatever their causes, are similar. Whether it is drug-induced fibrosis, glomerulonephritis, or hepatic cirrhosis, there is always an accumulation of extracellular matrix. SOD is effective in treating these types of fibrosis as well. Preliminary data suggest that such treatment prevents the occurrence of cancer.

Copernicus reexamined what certain Greeks had already stated: that the movement of the earth, the moon, and the planets is easier to understand if we rely on a simple hypothesis – the existence of a system with the sun, rather than the earth, as its center. Everything had already been said: that the moon revolves around the earth, that the earth is round, that the sun is the center of the system. Copernicus's genius consisted of assembling various dispersed ideas into a single logical, verifiable picture. However, it took over a century after the publication of his work for the scientific world to adopt this universal vision. When all is said and done, the same may be true for Warburg.

We are nearing the end of our voyage. Our objective is neither knowledge for its own sake nor the pleasure (albeit undeniable) of hypothesizing.

We do not have the means to continue this work on our own. Now, it is up to the institutions and the industry to take the ideas they find pertinent and apply them. To convince them to do so is what this book tries to accomplish. Glycolysis is a target for the pharmaceutical industry. This industry can develop both agonists and antagonists. The process might be fraught with difficulties. We now have the tools Otto Warburg was missing: molecular biology and molecular screening tools that make quick progress possible. If our hypothesis is correct, we should be able to advance rapidly, confirm his intuitions, and make good use of them. But let us hope that soon we will see glycolysis inhibitors in clinical trials, and that, for once, encouraging results will translate into undeniable benefits for our patients.

At the end of this journey, I feel as if we had simply rediscovered the wheel. It has been known for close to 100 years that fermentation causes cancer. There are several papers which suggest that blocking fermentation inhibits animal tumor growth. Some of these inhibitors, such as 2-deoxy-glucose, are already on the market and can be bought on the Internet. These molecules seem to be less toxic that standard chemotherapy. But, there are no human clinical trials proving or disproving their effectiveness. Why not?

Fashion may be part of the answer. Inhibition of glycolysis may be too simple, too rustic compared to the ever-increasing knowledge of molecular biology. Part of the literature on physical carcinogenesis, the extracellular nature of cancer or the importance of physical constraints, may have been forgotten. Or I may be simply wrong. The road to hell is paved with good intentions, and the road to science with hypotheses, often approximate and sometimes outright mistaken. Only time will sort out the good from the bad. Changing perspectives is a way to invent new therapeutic approaches. This is what we tried to achieve.